D0521367

THE HIGH FRONTIER

EXPLORING THE TROPICAL RAINFOREST CANOPY

MARK W. MOFFETT

Foreword by E. O. Wilson

HARVARD UNIVERSITY PRESS
CAMBRIDGE, MASSACHUSETTS, AND LONDON, ENGLAND
1993

PAGE 2: In Borneo, ecologist Tim Laman has climbed an equivalent of fifteen stories high in a tree that has a fruiting *Ficus stupenda* fig growing along its trunk. From there he shoots fishing lines over adjacent trees, where he will rig traps to monitor the rain of fig seeds at different distances from the parent plant (see page 166).

PAGE 5: A range of palm tree architectures, from James Orton, *The Andes and the Amazon* (New York: Harper & Brothers, 1876).

PAGES 6 AND 7: In Sri Lanka, a millimeter-long mottled gray "dust-speck" jumping spider (*Phyaces*) invades the nest of a much larger green jumping spider species to steal her eggs. She confuses the guarding mother by looking and acting like a wind-blown speck.

PAGES 10 AND 11: Ecologist Robin Foster identifies hundreds of species of trees growing in the 50 hectare (120 acre) plot on Barro Colorado Island, Panama (see page 37).

PAGE 13: Detail, from Henry Walter Bates, *The Naturalist on the River Amazons* (London: John Murray, 1864).

Library of Congress Cataloging-in-Publication Data
Moffett, Mark W.
The high frontier : exploring the tropical rainforest canopy / Mark W. Moffett ; foreword by E.O. Wilson.
p. cm.
Includes bibliographical references (p.) and index.
ISBN 0-674-39038-5 (cloth). — ISBN 0-674-39039-3 (paper)
1. Rain forest ecology. 2. Plant canopies. I. Title.
QH541.5.R27M64 1994
574.5′2642—dc20 93-16935
CIP

Printed in Italy.

Even where the trees were largest the sunshine penetrated, subdued by the foliage to exquisite greenish-golden tints, filling the wide lower spaces with tender half-lights, and faint blue-and-gray shadows . . . for what a roof that was above my head! Roof I call it, just as the poets in their poverty sometimes describe the infinite ethereal sky by that word; but it was no more roof-like and hindering to the soaring spirit than the higher clouds that float in changing forms and tints, and like the foliage chasten the intolerable noonday beams. How far above me seemed that leafy cloudland.

—W. H. Hudson
Green Mansions (1904)

To know the forest, we must study it in all aspects, as birds soaring above its roof, as earth-bound bipeds creeping slowly over its roots.

—Alexander F. Skutch
A Naturalist in Costa Rica (1971)

CONTENTS

PREFACE 8

FOREWORD *by E. O. Wilson* 12

1 TREE CLIMBING FOR GROWN-UPS 15
Scientists have just begun to explore the canopy's rich ecological tapestry.

2 SEEING THE FOREST FOR THE TREES 27
The diversity of rainforest tree communities is remarkable.

3 A PALACE OF MANY FLOORS 47
Tree architecture and spacing determine the structure of the canopy.

4 GARDENS IN THE SKY 65
Many canopy plants get by without ever touching the ground.

5 TAPPING THE GROUND 83
Some canopy plants go to great lengths to maintain connections with the ground.

6 INSECTS ON A RAMPAGE 99
Insects form the vast bulk of the rainforest's species diversity.

7 FURRED AND FEATHERED ON THE TOP OF THE WORLD 117
For large animals, canopy life represents a never-ending struggle against gravity.

8 A FLORAL SYMPHONY 135
The rhythm of canopy plant reproduction affects the diversity of tropical life.

9 TREETOP GAMES BETWEEN PLANTS AND ANIMALS 153
Most canopy plants entice animals to move their genes over the landscape.

10 A SCIENCE NEARS MATURITY 171
Rainforest canopy biology moves into the scientific mainstream.

ACKNOWLEDGMENTS 178
SELECTED REFERENCES AND NOTES 181
INDEX 190

PREFACE

How can one picture the vast hosts of these creatures peopling the soil and air, the roots, trunks, flowers, and fruits, and realize their metamorphoses, their habits, and the relations in which they stand towards the plants amongst which, or on which, they live? In short, how can we ever come to know the biology of this vast living world, which even the profoundest philosopher fails to grasp as a whole?

—ODOARDO BECCARI
Wanderings in the Great Forests of Borneo (1902)

If this book is unique, it's because it represents a photojournalistic overview of an entire scientific field—tropical rainforest canopy biology—that is in its infancy.

Trees overshadow my story more than they do most books about the rainforest. Trees also figure into the difficulties I've had in presenting my story. As the novelist John Fowles said of them, they "defeat viewfinder, drawing paper, canvas. They cannot be framed and words are futile, hopelessly too laborious." I can't expect to convey the vitality of rainforest trees and their residents with a book. But I can try to convey my awe and understanding, the result of months of firsthand experience.

From an early age I was awed by explorers. I devoured the tales of such early naturalists as Charles Darwin, Sir Henry Morton Stanley, Theodore Roosevelt, and William Beebe. I read Alfred R. Wallace's *The Malay Archipelago* three times, roaming the tropics with him in my mind. The story that impressed me most, however, was a fantasy—its author I don't recall. It featured a world covered with trees so fantastic that the characters walked upon the upper canopy as if it were solid ground. They descended to a layer of foliage below, a shadow world of other wonders. As they continued down through the canopy's strata, the tree trunks gradually became more immense, the darkness grew, and always there were new animals, plants, and adventures. I don't remember if they ever reached the ground; certainly it would have been anticlimactic.

Scanning neighborhood trees from atop the red maple in my family's backyard, I often thought of that forest, which, in contrast to the temperate forests that I knew, seemed unreal and yet so possible. But today, after many years of watching tropical creatures, I no longer consider it a fantasy. A tropical rainforest is a labyrinth of species and layers, one that forms over the centuries.

Exquisitely wrought stories of tropical forests can be found in books whose titles alone conjure up images: William Henry Hudson's *Green Mansions*, Alexander Skutch's *A Naturalist amid Tropical Splendor*, and Alex Shoumatoff's *The Rivers Amazon* are examples. The authors of these books went to extremes to see exotic places. Their writings describe the rainforest environment: extravagant trees, flowers, and animals; that mysterious atmosphere in which something splendid assuredly awaits—perhaps *there*, behind that trunk.

Admittedly, some authors have evoked forest imagery of a sharply different kind—as did Dante in the *Inferno:*

> that savage forest, dense and difficult,
> which even in recall renews my fear.

While I can appreciate the craftsmanship of Joseph Conrad's *Heart of Darkness* or Graham Greene's *Journey without Maps* (with his "dead forest"), my approach could never be of this negative sort. I love the rainforest. That's not to say I haven't spent miserably damp, mosquito-ridden days in the tropics, but I've had them in Wisconsin and Maine, too. Much of my time in the tropics has been boundlessly inspiring. Maybe the pulse of the organic world simply beats stronger for me there than anywhere else.

A lot is exciting about tropical rainforests, but I fear that the media's coverage of their demise has grown so prevalent that it has turned into white noise, easily ignored. A better educational approach is to show how spectacular the abiding forests can be. That is my approach in this book. Among those readers who relish nature, an urgency for saving the forests will, I hope, follow.

The most fantastical attribute of the rainforest is the canopy. In this book I want to capture how the canopy's environment is experienced by millions of species—occasionally humans. Innumerable books cover the life histories of specific animals. I devote little space to this, focusing instead on what makes canopy residents unique. Moreover, each chapter isn't simply about a type of plant or animal or interaction, but about an aspect of canopy ecology best illustrated by that plant, animal, or interaction. To sum up the theme of the book as a whole, I defer to the writings of Paul W. Richards, father of modern rainforest science:

> Although at first glance the jungle seems to be a totally disorganized chaos of superabundant greenery, closer examination shows that it has a rather definite structure. Tall trees, short trees, vines and epiphytes each have a specific role in the general scheme of things. And animal life in turn is well fitted into the overall architecture of the rain forest. In the jungle, in short, there is a place for everything—and everything remains pretty much in its place.

I describe—in more detail than you will find in other nontechnical books—how tropical canopies change vertically and horizontally, from day to day and from century to century, and what this tells us about trees as a substrate for life.

Current knowledge of the tropical canopy centers on the Americas, but, because no two forests are the same, I have tried to incorporate information from elsewhere when it exists. Africa is underrepresented, and, as of 1993, the canopies of Madagascar remain virtually untouched by tree-climbing scientists.

Today it is still possible to grasp rainforest canopy biology by talking with the small group of experimental scientists now active in the field. I decided to seek out as many of these frontier scientists as I could, to see what drove them into this unfamiliar and at times perilous environment, what methods they chose for getting about, and, most important, what gifts of knowledge they have given us from the tops of tropical trees. Between March 1989 and December 1992, I visited thirty or so rainforest canopy specialists at seventeen major research sites worldwide. This book is dedicated to them, the canopy explorers. —*M. W. M.*, May 1993

FOREWORD

It is for me a stunning fact that while the physical surface of the earth has been thoroughly explored, so that virtually every hilltop, tributary, and submarine mount has been mapped and named, the living world remains largely unknown. As few as ten percent of the species of insects and other invertebrate animals have been discovered and given scientific names. Bacteria and fungi are even less well known, with less than five percent named to date. Even new kinds of mammals, birds, and flowering plants turn up each year.

The richest habitats on earth are the tropical rainforests. Although they cover only six percent of the land surface, they contain more than half the species of plants and animals on earth. They are also, together with the deep sea floor, the least examined of the biotically rich environments. And within the rainforests, the canopy is the remote outland, the final frontier. It is another continent of life, in William Beebe's vivid imagery, that exists not upon the earth, but one to two hundred feet above it. Capturing most of the sunlight, conducting most of the forest's photosynthesis, and containing a majority of the species of animals and plants, the canopy has until recently been physically almost out of reach. The tree boles are difficult to climb, rising largely branchless for most of the way up. They are smooth or spiny in surface texture and often exude sticky latex when scratched. Climbers who make the top often encounter swarms of ants and vicious wasps, and they must make their way in and around impenetrable and dangerously wobbly mats of epiphytes.

Only within the past twenty years have biologists, mostly young and athletic, devised ways to get into the canopy and stay there long enough to explore it systematically. Their efforts, chronicled expertly in this well-written and beautifully illustrated book by Mark Moffett, have just begun. Beebe's aerial continent is now opening up. It composes an area approximately equal to that of the contiguous forty-eight United States, waiting to be climbed, walked, and mapped hectare by hectare; and within it live millions of organisms still unknown to science.

As Moffett explains, the *idea* of the rainforest canopy has captured the European and American imagination for nearly two centuries. It sustains the archetype of an unending land, still untrammeled and paradisaical. To the dream has now been added modern climbing techniques and the measurements, classification schemes, and syntheses of modern science. The new era of exploration will not corrupt the rainforest but will add immeasurably to its understanding and appreciation. Science, in short, will enrich the dream.

No pilgrim is better equipped to report these pioneering years than Mark Moffett. He spent two years in the rainforests of Asia for his doctoral research, and has since personally explored almost all the other major tropical forests around the world, traveling long distances by boat and on foot along the shadowed trails and climbing up ropes, towers, and ladders into the canopy. I had the pleasure (and anxiety) of watching him teeter on the top of the canopy tower of Panama's Barro Colorado Island, strapped loosely to projecting meteorological instruments, to get photographs at just the angle he sought. He is the personal friend of nearly every important canopy researcher in the world, accompanying many of them into the field. Moffett is tireless, thrives in sweat-soaked clothing on torrid afternoons when others fold, and is seemingly immune to tropical diseases. His book is a tour de force of scientific journalism, as well as the visual and literary expression of an experienced and deeply caring naturalist. — *Edward O. Wilson,* Frank B. Baird, Jr., Professor of Science, Harvard University

CHAPTER ONE

TREE CLIMBING FOR GROWN-UPS

Growing up in the midwestern United States I knew trees well. I looped from one bare branch to the next in the backyard red maple with, I believed, the speed and grace of a monkey making its rounds. Like Kipling's Mowgli, I had the position and strength of each branch memorized. I learned how to rest my body comfortably among the orderly boughs in order to have a clear view of my mother, small as an ant, tending the garden below. The branches I favored became burnished from repeated scuffings. In time I identified with the monkey's world. I grew up to be a zoologist.

Animals have always been my passion, especially maligned creatures like snakes and ants. Whereas authorities on primates and other big social animals take months or years to gather data, with ants I can house a society inside a shoe-box-size container and unravel some of its mysteries within a week.

In college I befriended experts on tropical reptiles, butterflies, and beetles at the Milwaukee Public Museum and became an assistant for their research. My first venture afield—and my first experiences with the richness of tropical canopies—was with Max Nickerson, a childhood hero of mine. I first met Max at the charter meeting of the Wisconsin Herpetological Society when I was twelve. He captivated me with his dashing appearance, booming laugh, and stories of adventure. As a youth he had lost a finger after a snakebite; he knew how to catch crocodiles by hand.

When I traveled to Costa Rica with his research team in 1976, I was living a dream. By day we used radiotelemetry to track gaudy "Jesus Christ lizards" (they run over water) up into the canopy along a rainforest-lined stream. At night we sought famously long parrot snakes high in the same tree crowns. I was known for quick hands and was elected snake catcher. When a snake was sighted by flashlight, I had the job of standing knee-deep in the cool murky waters below while biologists on the bank maneuvered a pole with a hooked tip into the tree. I stared hard as the pole wavered above me in the headlamp beams, raising my arms as the hook inched below the unsuspecting serpent, who was looped among the twigs. Suddenly the pole was thrust upward to snag a coil. The dislodged snake spun off the hook into the damp atmosphere. Time slowed. The snake shot through light beams as it descended, one moment invisible against the inky canopy, now—fleetingly—a brilliant green twirling strip. My hands predicted its trajectory faster than my mind could grasp. In the two times (out of three) I succeeded, it seemed miraculous that one hand cradled the fall of the lithe body while the other seized the snake behind its parrot-beaked head where it couldn't bite.

As the others gathered round I marveled at the snake's interminable length and wondered why life in tropical treetops necessitates such an elongated form. Max fed the serpent a wax-coated radio transmitter and we released it on the same tree, plotting its ascent with the beeps of our telemetry device.

With these first tropical experiences I was hooked. Within a year of beginning Ph.D. studies at Harvard I took off to tropical Asia, where I tracked down my childhood preference—exotic ants—roaming for twenty-eight months through forests from Nepal to Taiwan.

Early in my career as a zoologist I kept my eyes down. Rainforest trees cloaked me in shadows, immense and daunting—and as unreachable as Venus. Once in a while I caught a glimpse of a flashy bird or mammal, or heard one of their plaintive cries. For the most part, though, I remained oblivious to the canopy. I realized only gradually that I would miss out on most ants and other favorite creatures unless I got into the trees.

Now I faced a classic problem of rainforest exploration: I felt landlocked, the canopy ocean beyond reach. I resorted to the strategies of my predecessors. I scrambled across the shattered limbs of trees thrown over by wind, clutching prized insects in my hands. In Borneo, tribal people, slighter and more agile than I, clambered up trees to harvest mangoes and other forest fruits. I hired two of them to search

PAGE 14: The author climbing a rope into the canopy in Monteverde, Costa Rica. ***(Photo: John T. Longino)***

PAGE 15: Weaver ants patrol atop a tree leaf while the jumping spider *Myrmarachne,* hiding beneath it, mimics the ants with her fake eyespot and forelegs held like antennae. She tricks predators that avoid the noxious ants but must keep away from the ants herself.

LEFT: Like other arboreal snakes, the parrot snake (*Leptophis depressirostris*) can stiffen its body to bridge gaps in vegetation up to half its length. It sneaks up on agile prey such as frogs and bites them, injecting venom through rear teeth.

BOTTOM: Red-eyed tree frogs (*Agalychnis callidryas*) stroll the canopy but also visit the ground to breed on leaves overhanging pools in the rainforest (Panama).

for insects, but they lacked trained eyes. Brandishing a machete I felled a tree harboring armored ants in a Venezuelan logging site. I'd have preferred to climb and observe the ants while leaving the canopy unmolested. I recalled the despair of the Englishman Richard Spruce, who documented Amazonian flora in the mid-1800s, occasionally chopping down trees because of the scarcity of experienced tree climbers in that region:

> It was a long time before I could overcome a feeling of compunction at having to destroy a magnificent tree, perhaps centuries old, merely for the sake of gathering its flowers. . . . In the same way I suppose a zoologist stifles his qualms of conscience at killing a noble bird or quadruped merely for the sake of its skin and bones.

Many people study felled trees, but I wanted to know how creatures survive within a living, respiring tree. Despairing of my earthbound existence, I scaled trees to observe jumping spiders in Sri Lanka. But climbing without safety aids limited me to small trees with sturdy branches. Besides, the hardships of trekking to the nearest hospital lurked in the back of my mind.

Then I talked to Jack Longino, a professor at Evergreen State College in Washington. Jack was an ant enthusiast, too, but he knew how to reach canopy species by rope, using gadgetry developed for rock-climbing. He offered to teach this technique to Harvard graduate student John Tobin and me at Costa Rica's Monteverde Cloud Forest Reserve.

What makes a tropical rainforest? At first glance rainforests look like tangles of vegetation, but just as familiarity with buildings gives one an appreciation of individual architectural styles, long experience allows one to recognize the rainforest's species diversity and structural logic. Painstaking floral surveys in New World rainforests have shown that, even excluding their trees, rainforests rank as the most diverse of biological communities—with more herbs, shrubs, and canopy plants than any other ecosystem.

Rainforests vary from region to region and locally with changes in altitude, soil, topography, wind, and other factors. New World rainforests have tightly

The lowland tropical rainforests of Ecuador are among the world's richest in animal and plant species.

packed crowns interwoven with vines and filled with other canopy plants. Equivalent Asian forests have an abundance of towering trees with narrow crowns that resemble giant lollipops. Branches may be loaded with arboreal plants, but seldom in densities reached in the New World; lianas—woody vines—are notably scarce. In Africa, trees tend to be short, with umbrella-shaped crowns and few canopy plants other than vines. Canopies there are generally broken by wide spaces between crowns, especially where such large mammals as elephants crisscross the earth.

Even visitors who never consciously dissect such attributes may pick up on shifts from forest to forest. Differences can be as elusive as faint odors, the timbre of ambient noise, or the forest palette, the latter perhaps more a result of local nuances in light than in colors intrinsic to the forest. Aficionados revel in a rainforest's sensory opulence. The moment I enter a forest, wafting scents (conceivably from hidden canopy flowers) and delicate hues awaken memories of similar places, like the tumult of images that well up when I enter an attic filled with objects long neglected.

In addition to these subtle cues, obvious peculiarities mark some forests: the eerie booming of gibbons fills Southeast Asian forests; pineapple-top bromeliads cram Latin American trees.

Altitude plays a central role in rainforest variation, although factors such as rainfall and humidity also affect the range of forest types. In lowland rainforests unscarred by human hands, tree trunks tower like colonnades, supporting crowns that soar eighty to two hundred feet high. Branching begins high. Mammoth trees shoot past less imposing neighbors; even their lowest limbs may be impossible to make out through the foliage below. Canopy plants grow patchily on branches. Walking can be easy in unspoiled forests: only a smattering of young trees and vines impedes the way. Mists disperse each morning. Rain arrives in torrents between intervals of intense sunshine.

Forests on the slopes of mountains at first seem little different from the lowland rainforests. But starting between 2,500 and 9,000 feet on moist slopes

Cloud forests can occur both at sites that are shrouded in clouds but receive relatively little precipitation and at sites that are drenched by showers every day. New Guinea has extensive cloud forests at unusually high altitudes, often starting above 7,000 feet.

one enters cloud forest. Mists there linger during the day and higher up the forest may be perpetually engulfed by clouds. Plants change in response to the decline in solar radiation and temperature. Trees are less imposing. The canopy, reaching fifty to one hundred feet, becomes simpler, with fewer layers and a more even roof. Arboreal plants carpet the branches and trunks virtually wall-to-wall. This plush greenery sponges up sound, making animals hard to locate.

On wind-raked slopes at cloud-forest elevations or higher, leaves shrink; trees become stunted and contorted, branching closer to earth. This is elfin forest. Visitors push through a confusion of continuously soaked, gnarled limbs and stumble over exposed roots. In many elfin forests you can see over the canopy while standing on the ground. Arboreal plants drape the trees so thickly that it is hard to tell which leaves belong to a tree and which are parts of intruders on it. It can be impossible to walk. Perhaps this is botanical heaven—or is it hell?

Sometimes it is hard to delimit these forest types and, because of centuries of global human intrusion, to specify the properties of truly "pristine" forest. Depending on how far from settlements travelers are willing to go, one person's virgin rainforest may be another's second-rate grove.

Moreover, some specialists restrict the general term "tropical rainforest" to a specific climate based on annual levels of rain and humidity, regardless of elevation. Others, focused on the communities themselves, include places where rain, although marginal (as low as fifty inches per year), falls so evenly that most trees stay evergreen, as in parts of India. Other tropical ecosystems, such as those that are less moist, more seasonal, or both, are invariably excluded.

Most of this book, except for visits to mountainous tropical regions in Costa Rica, New Guinea, and Colombia, will focus on the tropical lowland rainforests in Asia, Africa, Central America, and South America. Hereafter, this is what I'll mean by "rainforests."

In March of 1989, I arrived at Monteverde Cloud Forest Reserve in Costa Rica feeling nervous, confiding in my companion, John Tobin, about nightmares of falling from the ropes Jack Longino would teach us to climb.

Dauntless Jack didn't allay my fears. His idea of instruction turned out to be minimal. As John and I struggled with a puzzling jumble of waist harnesses, clips, ascenders, metal loops, and safety cords—assembled with unnerving haste by the harried clerk at the store where I had bought the stuff—Jack provided a monologue about the lethal mistakes we could make. After tinkering with our equipment for a moment, Jack pointed up.

"Now?" I said incredulously.

"Now!" he answered.

I clipped on to a climbing rope that extended into the tree. Five minutes later I was still bouncing unceremoniously on the muddy ground as I tried to pull myself up the rope. Ten minutes more and my muddled first attempt at an ascent had gotten me a few yards from earth. Meanwhile, my voice had climbed an octave because of a badly positioned loop in my harness.

Although by my later standards the tree we chose was small (perhaps eighty feet tall), climbing it surpassed the backyard experiences of my childhood. Dirt and debris coated me at once. Lack of an easily detected order confused me; encircling greenery created a visual cacophony. Each branch swayed independently—including the one I straddled—and the absence of a stable reference point threw me constantly out of equilibrium. A path worn along one branch marked a monkey's route. The world of that monkey was more foreign to me than I could have ever imagined.

John and I clutched the tree limbs so hard our muscles soon ached. Would some curious animal gnaw through our rope when we were not looking? Were our safety lines securely tied? Suddenly, the possibility of bumping into a hornet nest became a matter of life and death. Wasps and bees are always a concern in a rainforest, and it is hard enough to sprint away from them on the ground. In the trees we had to avoid meeting any in the first place and we kept alert for them at every turn.

Barely audible shouts of encouragement from our earthbound companions, veiled beneath reced-

Elfin forest trees tend to have stout branches and twigs to resist windsnap, and the trees themselves are stunted by the high winds, cool temperatures, and poor, leached soils (Mount Kinabalu, Sabah, Borneo, altitude 9,500 feet).

ing floors of foliage, simply confirmed the absurdity of our height. We had heard too many rumors of malicious beasts, failed equipment, and cracking boughs to simply relax and take in the view. Not all dangers were hearsay. There have been serious canopy injuries. In Venezuela a few years back a student died when his rope snapped. But despite our qualms, the treetops thrilled us and overloaded our senses.

Still giddy at the end of the day, John and I counted our bruises, brushed leaves from our hair, and washed the soil and bark from our skin. Then we joined Jack and his wife, biologist Nalini Nadkarni, at their Monteverde cabin for heaps of rice and chicken. Filled with bravado from the climb, we talked adventure. Part of my exhilaration, I admitted, came from the exertion that had been required to reach the treetops; though I had been concerned about the height, I had somehow imagined the climb would be effortless.

All of us admitted to a childhood fascination with trees. "This," Nalini said, gesturing at a mound of ropes and gear, "is tree climbing for grown-ups!"

Jack and Nalini told us about research in tropical treetops. A few years ago you could count the intrepid canopy specialists on your fingertips; in 1993 you'd need six hands. In comparison, hundreds of rainforest biologists worked terrestrially worldwide. Even so, few arboreal biologists had met each other: most worked alone.

Technology to assist biologists who wanted to work in the canopy advanced more slowly than comparable equipment for undersea divers, which,

Moss forest, occurring in some regions at the transition between lowland rainforest and cloud forest, is perpetually soggy with tall moss-sheathed trees (Papua New Guinea, altitude 5,100 feet).

FACING PAGE: In a tree swathed with canopy plants at Monteverde, Costa Rica, Jack Longino and Nalini Nadkarni take a lunch break during a day of studying ants and plants, respectively.

by the 1940s, had already made Jacques Cousteau's expeditions possible. As biologist Marston Bates described in 1960:

> With the invention of the Aqualung and similar devices man has gained a freedom in the seas (at least the top hundred feet or so) that has no counterpart in the forest; but even before this I think diving was easier for him than climbing.

Still, climbing methods had progressed rapidly beyond reliance on local climbers and felled trees, which had been prevalent since Darwin's day. Thirty years ago, the nailing or chopping of steps on trunks was common—though many tropical woods are so dense that nails can't be driven into them. They are squeezed back out into your face by the wood.

When a friend of mine, Roger Swain, pursued his Harvard Ph.D. on Brazilian ants in the 1970s, he walked up tree trunks with spiked shoes. This often worked fine. Then there were trees that he climbed happily, only to be met on the descent with waterfalls of latex pouring from the pierced bark. Removing the viscous stuff from his hair took a gasoline shampoo.

Techniques like these have been superseded by others less destructive to trees. Many of the "new" methods, like platforms and ropes, actually had first been tried decades before. The development of stronger, lighter, cheaper, and more reliable construction materials simply made such approaches more practical.

The transition to new technologies began in the mid-1970s, when entrepreneurs Donald Perry and John Williams first applied rope techniques perfected for mountaineering to tropical tree climbing. They touted the methods both to scientists and the public. Perry earned a Ph.D. in biology from the University of California at Los Angeles and an award for mettle and innovation from the Rolex watch company, but his flamboyant, often temperamental personality led to clashes with scientific colleagues.

Still, as a result of the pair's efforts, rope-climbing earned acceptability as a research tool—and canopy research began to blossom. Today many scientists use ropes. Others prefer cranes, ladders, walkways, booms, platforms, cherry pickers, towers, or tribal methods. Even with the modern approaches, canopy devotees must push themselves to the limit, as Roman Dial couldn't resist spelling out in his doctoral dissertation on Puerto Rican lizards:

> Nor can I describe all the hardships: how the trees trashed three pairs of shoes and four pairs of pants, how I lost two cameras . . . and a Swiss army knife (dropped from great height and swallowed by the jungle), how ants in the eyes and scorpions in the crotches (of trees) make tree climbing downright uncomfortable at times.

As that evening's musings with Jack, John, and Nalini continued, I grew aware that a profession—rainforest canopy biology—was being born. As yet no university offered a canopy course, and to this day studies remain so scarce that most canopy biologists are solitary masters of whole disciplines. But to me, this is the essence of a pioneer spirit in science: to seek a novel topic and claim it, temporarily, as one's own. You share the perspective of the explorer who, in mapping a terra incognita, pinpoints regions that offer extravagant rewards. Canopy explorers, I came to see, rank as frontier people both physically and intellectually.

Purists claim that the term "canopy" applies to the thin veneer of leaves struck by direct sunlight—the region of maximum photosynthesis, comparable to the ocean surface. To other biologists the canopy encompasses the crowns of these uppermost trees and the fan of limbs beneath the foliage. I use the term broadly to embrace the whole vegetational ocean beyond easy reach of ground dwellers, say from twenty or thirty feet up. This covers all the poorly known strata—including the middle (understory) levels.

A major cornerstone of tropical diversity are the beetles, new species of which are found every day. In some places the well-known golden scarab (*Plusiotis*) comes in equally impressive silver (Monteverde, Costa Rica).

The alien nature of this canopy region is depicted elegantly in Henry Tomlinson's 1912 travelogue *The Sea and the Jungle:*

> The forest of the Amazons is not merely trees and shrubs. It is not land. It is another element. Its inhabitants are arborean; they have been fashioned for life in that medium as fishes to the sea and birds to the air. Its green apparition is persistent, as the sky is and the ocean. In months of travel it is the horizon which the traveler cannot reach.

The canopy is to tropical biologists what the sea

once was to marine explorers. Just as the ocean first lured Cousteau in his untested Aqualung, treetops draw us despite perils and discomforts. "From this day forward we would swim in an alien land," Cousteau mused. "Across miles of country no man had known, free, and level, with our flesh feeling what fish scales know." A major distinction between hazarding the ocean like a fish and hazarding the canopy like a monkey is that we normally forge our way into one from above and into the other from below.

The water of the rainforest reef is the balmy air itself, through which minute organisms drift like plankton while larger ones fly and leap from one part of the tree scaffolding to another. Life extends from the crowns into the nether regions, an arena spanning in some places the height of a twenty-story building. Below, as on the ocean floor, lies a more stagnant, somber world, fed by a rain of dust, feces, branches, leaves, and animal corpses from above. The splendor we perceive derives largely from the canopy, with our view of sunlight cascading through tiers of foliage. The trees bridge it all, shifting most organic matter back into their crowns as soon as it becomes available (by decomposition) to their roots.

Jack, Nalini, John, and I enjoyed a starry Monteverde night. Eventually talk drifted from climbing and canopy biology to biodiversity—the diversity of living things. I had reservations about the emphasis placed on this recent catchword by most scientists and reporters, who presented it as little more than the enumeration of species. The notion of opulent tropical floras and faunas may have captivated people for centuries, but I argued that to sustain momentum for conservation, and to draw new intellects eager for challenges, biodiversity would have to mature as a subject. Not only should it encompass many of the provocative concepts from ecology and evolutionary biology, it must radically expand upon them.

The greatest impetus for conservation, I realized, stared us in the face. It was not biodiversity as such, but rainforest canopy biology. Much of earth's biodiversity adorns the crowns of rainforest trees, and in combination those species form a community of unparalleled complexity. Our reflections that evening on the grand pageant of treetop interactions transformed my thinking. Once, I'd envisioned the canopy as those parts of trees that just happen to be up, but it's far more than that: it is an ecosystem in its own right, a tapestry of interconnected species spread over thousands of square miles. As the explorer Alexander von Humboldt wrote in 1815 describing the tallest tropical trees, the canopy is, in effect, "a forest above a forest."

Treetop and deep-ocean explorer William Beebe predicted the canopy's profound biological significance back in 1917. His insightful writings could serve as a rallying cry for canopy biologists:

> Yet another continent of life remains to be discovered, not upon the earth, but one to two hundred feet above it, extending over thousands of square miles. . . . There awaits a rich harvest for the naturalist who overcomes the obstacles—gravitation, ants, thorns, rotten trunks—and mounts to the summits of the jungle trees.

The web of species interactions in this "lost" continent intrigues me most. We may contend that a liverwort or ant has value based on aesthetics, chemistry, or the information in its DNA, but the significance of a species to life's web overshadows all else. The liverwort depends on certain species; other species depend on it. Remove the liverwort and part of the web collapses. To promote the survival of a scarlet macaw over a liverwort, then, is to promote Michelangelo's paintings over the Sistine Chapel; they must be recognized as part and parcel of the same structure. Canopy biologists examine the most exquisite tapestry of life on earth, which may be our most precious biological resource.

CHAPTER TWO

SEEING THE FOREST FOR THE TREES

I stepped from the wood flooring of the entrance gazebo onto a plank footpath hung in space. The solid thunk of the floor under my feet was replaced by creak and sway. Already the earth lay fifteen feet down. I advanced warily, adjusting to bounces that came with my footfalls while sliding my hands along the cables on either side. The suspended path was level, but because the ground sloped away below me I was heading toward the upper canopy. Ten yards forward and understory crowns greeted me. Ten more and I was all the way up among the topmost trees. In one painless, horizontal stroll, I had made a journey that normally would have required a vertical ascent equivalent to scaling a twelve-story building.

The Malaysian state of Sabah in northeastern Borneo is on few lists of must-see places. Yet it is here, at Poring Hot Springs in the foothills of Mount Kinabalu, that anyone who wishes can enter the rainforest canopy world. The six hundred feet of walkway I had begun to traverse that day has been open to the public since its completion by the Malaysian Park Service in 1990.

The first leg of the walkway ended at a platform around a tree trunk. German, Dutch, Asian, and American tourists were congregated there. They had come with a minimum of expectations, perhaps to see birds or to experience height. Once they had overcome their nervousness, they were entranced by sights they had never anticipated. Golden thorny ants sauntered along the railing. A specimen of the world's largest orchid emblazoned one tree.

But the tourists also expressed awe at the tree giants themselves. A canopy is, after all, a vegetable construct; its tree foundation is central to almost every question you can ask about it. Indeed, it is but a slight overstatement to describe tropical rainforest ecology as the study of the relationships of organisms to trees. Not surprisingly, most canopy research has been grounded in part on understanding the trees: the number of tree species; how those species compete for room on the ground or within the canopy; the trees' role in cycling nutrients and energy through the forest.

Visitors to rainforests are often surprised by how fascinating the plant life is, something the Norwegian explorer Carl Block related in his 1881 book *The Headhunters of Borneo*:

> It was a long time before we saw bird or animal . . . but I was lost in admiration over the magnificent trees, and their beautiful variety of foliage, and the orchids and creepers with which they were covered, regretting all the time that I was no botanist, and that the true meaning of much that I saw was lost upon me.

No matter what canopy explorers claim as their real vocation, most end up as at least part-time botanists. The very act of entering the canopy requires a respect for trees. I have been in trees where every branch invites me to rest, and in others whose angled boughs forever jam me into tight corners or whose wood threatens to snap at every turn.

As the tourists and I admired the trees, John Beaman, an authority on Sabah's plants from Michigan State University at East Lansing, joined us on the walkway. John and I had talked before of the remarkably diverse flora of these mountain slopes. John wasn't a climber. He admitted to retrieving canopy specimens the traditional way: at treefalls. A walkway is of minimal value for finding new plants, he pointed out; as it is fixed in space, one can reach only a limited number of specimens. But he told me of scientists like American conservationist Illar Muul for whom walkways are invaluable. A walkway designer, Muul had a hand in developing this one and was in China at that time starting another. The traps for small canopy mammals that Muul used in his ecology research were propped within the trees all around us.

A moment after downplaying the importance of walkways for his own work, John found a tree that he did not recognize. He pointed out the top of its crown, twenty feet below the walkway. It was strewn with brown fruit that would tempt any botanist. Fruits offer traits often needed for plant identification, but I reminded him that it would be impossible to obtain a sample from where we stood.

A forest guard interrupted us. Not to worry! He

PAGE 26: Poring Hot Springs, in Sabah, Borneo, has two suspended walkway systems, one open to the public and the other available to scientists for canopy research.

PAGE 27: Made of wood, high-tensile ropes, and steel cables, the Sabah walkways diverge in different directions. Platforms around trees are stable places from which to observe canopy life.

The foliage textures and crown shapes of tropical trees in a New Guinea cloud forest reflect the canopy's structural and chemical variety.

would send for Dominic, one of the three tree climbers who had built the walkway and whose agility was gained from supplying local kampongs, or villages, with rambutans and other rainforest tree fruits.

Minutes later a Malaysian in his late teens, dressed neatly in Keds, jeans, and a button-down shirt, met us on the bouncing path. Deftly tying a short rope to a tree trunk, he tossed it over the netting that flanked the walkway and virtually threw himself after it. He seized the rope with his bare hands and descended in swift jerks hand under hand until he was positioned by the tree's crown. Ignoring the hundred-foot drop to the streambed below, he spun on the rope, feet thrashing. One hand holding his weight, he reached out at the foliage with the other, grabbed it, and pulled himself into the crown of the mystery tree.

Clambering among the tree branches, our climber snapped off three fruit-bearing leaf sprays and tied them to his rope. John Beaman pulled them up. "Sapotaceae," he said, looking them over critically. "The same plant family as chewing gum." So, chewing gum comes from a rainforest tree? "Of course!"

Dominic still had to get back. When he swung into space again the loose end of the rope became scrambled around his shoes. Unraveling it, he lost his grip with his feet twice in quick succession, dropping several feet before seizing the rope near its hanging end. Nonplussed, he dangled, thinking. He darted out of sight into the tree's crown, reappearing with shoes tied behind his back. He snaked up to the walkway in thirty seconds, rope clenched between bare toes. To traditional village tree climbers, I thought, even shoes are a technical distraction.

Asrab, an Indonesian tree climber, works for biologists in the southern (Indonesian) part of Borneo.

Variation in Tree Communities

More than any other factor, the types of trees at a site affect the life that will be found in the canopy. So to comprehend canopy diversity, we need to know the geographical distributions of different trees.

While any number of tree families might do, a family called the Dipterocarpaceae provides a good example. Dipterocarps are as widespread in Sabah and elsewhere in tropical Asia as oaks are in North America. The name, derived from classical Greek, means "two-winged fruit," a trait of many species. Certain dipterocarps, including such valued hardwood trees as the *Shorea* and *Dipterocarpus* genera, are huge, with compact crowns. Lowland forests such as the one at Sabah's walkway are impressive in stature and distinctive in appearance largely as a result of dipterocarps. In Malaysia, where I first saw their immense, widely spaced trunks, I once fancied them as the pillared legs of some leviathan that could squash a brontosaurus. As Norman Myers, a leading conservationist and ecologist, put it:

> Dipterocarps are tall trees, and their trunks, if slender by some standards, are impressive growths, straight and unencumbered for dozens of meters, soaring sheer from the ground to the canopy that is so far overhead that it seems to belong to a different world.

Imagine riding in a small plane over the Malay Peninsula. The rumpled canopy blanket flows by, static except when the engines set monkeys and squirrels leaping and hornbills into flight. The cauliflower-like humps of dipterocarp crowns stud this surface, so that it resembles, to a zoologist's mind, the bumpy skin of one of the geckoes hidden within the trees below.

As the plane heads southeast into Indonesia, the humps eventually grow sparse. We have crossed a geographic frontier: Wallace's Line, named for its discoverer, nineteenth-century naturalist Alfred Russel Wallace. This line, which demarcates a shift to a more Australian type of flora and fauna, passes between Bali and Lombok, continues through the Makasar Strait separating Borneo and Sulawesi, then curls below the Philippines. By the time we reach New Guinea, dipterocarps are the rare exception; other tree families have replaced them. The canopy surface could be confused with that of the distant Amazon, where dipterocarps are insignificant.

If we could distinguish individual species of dipterocarp from the air (a feat requiring superhuman vision), we would have witnessed an astonishing decline in dipterocarp species on the flight. Borneo harbors 267 species, but only seven grow immediately to the east of Wallace's Line on the cross-shaped island of Sulawesi. Furthermore, species overlap in a patchwork: some range over several islands; others are confined to one island or part of an island. Each forest offers a distinctive assortment of dipterocarps.

Dipterocarp abundance and species diversity also decline as one heads from Malaysia westward to the Indian subcontinent. The family regains importance on the island of Sri Lanka, near India's tip. Dipterocarps there are more closely allied to those of distant Malaysia than to any in India. Studying such distributions is the lifework of biogeographers, who puzzle out how organisms spread around the globe.

No other family dominates a broad tropical rainforest region to the extent that dipterocarps dominate Southeast Asia, although trees of the legume (bean) family come closest in Africa and the Americas. Yet the variation in the species richness and abundance of dipterocarps from site to site and from region to region differs only in degree from that for hundreds of other groups of rainforest trees. Continents or regions within them share many plant families but fewer genera and even fewer species. Consequently, every rainforest around the globe has a distinct tree community, which in turn yields a structurally and chemically unique landscape for arboreal organisms.

THE RARITY OF TREE SPECIES

While I was canoeing across Ecuador's wide Napo river, I noticed that the rainforest on the far bank at first seemed a delicate crust on the land, like mold on cheese. Trunks of trees came into focus as pale slivers throwing wisps of green skyward, and slowly the canopy expanded and solidified into a variegated carpet beneath the swollen gray clouds. Branches and vines soon became visible.

Once I was this close, the analogy with a mold broke down. Under magnification, a mold typically resolves as a single form that repeats monotonously. Yet in this forest each tree had a form distinct from its neighbors, a unique way of arranging its finery. Where were there two of the same species? Alfred Russel Wallace reported his bewilderment with the rarity of many Asian trees in 1869:

> If the traveler notices a particular species and wishes to find more like it, he may often turn his eyes in vain in every direction. Trees of varied forms, dimensions and colours are all around him, but he rarely sees any one of them repeated. Time after time he goes towards a tree which looks like the one he seeks, but a closer examination proves it to be distinct. He may at length, perhaps, meet with a second specimen half a mile off, or may fail altogether, till on another occasion he stumbles on one by accident.

Given statements like this about rainforest diversity, the Congo basin is surprising. One moment there is the expected profusion of tropical trees and canopy tiers; then the understory opens up to a clear view of trunks receding in the distance, each golden with fine, flaky bark. The foliage overhead is a monotone; the litter underfoot is uniform with the leathery leaflets of a single tree species. At the right time of the year, red petals or massive disk-shaped seeds fleck the ground.

This is one of the stands of *Gilbertiodendron dewevrei* trees that occupy vast portions of the basin. These stands alternate abruptly with rainforest swaths rich in tree species. *Gilbertiodendron* stands are a dramatic example of a monodominant tropical rainforest; there are less extensive cases around the world in places like Trinidad, Australia, Uganda,

PAGES 32–33: In the tropical rainforest near Manaus, Brazil, climax (old growth) forest shifts abruptly to an assortment of pioneer (early successional) tree species, including *Cecropia* and palms. The forest at this site may once have been felled for pasture.

Guyana, and Borneo—each involving a different tree species.

In monodominant forests, not only are single species (or a few related ones) numerous, but they dominate an area ranging from a small patch to a whole region. Sometimes a species is in a pure stand; but in cases like the *Gilbertiodendron* forests of the Congo, the same set of tree species found in nearby "diverse" forests do still exist in the canopy of the monodominant forest, but one species has become exceedingly common and all other trees exceedingly rare. The dominant species gives each monodominant forest a unique feel: the repetition of form, texture, and color in the trunks and crowns is reminiscent of temperate forests, but the trees themselves may appear more than a little outlandish.

The sites I've described from Ecuador and the Congo lie at two extremes in a continuum, from forests in which virtually every tree is different from its neighbor, to forests in which virtually every tree is the same as its neighbor. Not only, then, do rainforest trees vary from region to region, but the diversity of the trees varies in different places, as do the local distributions of the species.

Ecologists are trying to understand how tropical trees are distributed both locally and regionally. They agree that distribution is influenced by physical conditions and by competition among trees, young and old, for space, but they bicker over the relative significance of these factors. In truth, all may be significant—it is a matter of the scale at which you look at a forest. Ecologists who work in large tranquil sites for a few years come to different conclusions about forest dynamics than ecologists focused on small windswept locales over decades.

Sri Lanka's *Shorea gardneri*, a dipterocarp tree, flourishes in congested monodominant stands that accentuate the characteristic bumpiness of their crowns. *(Photo: Darlyne Murawski)*

Physical Environment

Like all organisms, trees grow under restricted conditions; each species has certain habitat requirements. Over geological time the tropical environment has been in flux. In dry epochs, many scientists believe, conditions worsened; surviving rainforest tree species contracted to isolated pockets of favorable habitat called refugia. Populations evolved independently during the periods of isolation in refugia, which in many cases eventually led to formation of new species. When rainfall and humidity increased, the ranges of the trees (and the canopy organisms harbored by them) expanded again.

The landscape can bring such species expansions to a standstill. The ranges of many dipterocarps stop short at rivers. Perhaps these trees are confined by the meager dispersal of their winged fruit, which whirl downward like overweight maple seeds.

But species ranges need not be determined by anything so obvious as a river. Though the weather may seem invariant in rainforests, climate differs enough from place to place to affect species. For example, most rainforests show some seasonality in precipitation, but the area encompassing Sumatra, Borneo, and the Malay Peninsula experiences nonseasonal rainfall. At a rainforest site under survey in Borneo as this book goes to press, botanists have recorded 995 tree species in twelve acres, an area one can circle on foot in a leisurely twenty minutes; hundreds more occur in adjacent forest. Few regions on earth harbor such a bounty of trees. An equivalent area to the north, in Thailand, where rainfall is relatively seasonal, may hold a fraction of that number of species, and most of these would be distinct from those to the south. This would still be more species than are indigenous to all of eastern North America, an area at least one hundred million times as large.

Water affects tree distribution in other ways. In the Amazon basin, seasonal runoff of water from the vast river drainage creates widespread flooding. The severity of flooding produces a mosaic of trees tolerant of different degrees of inundation. In hard-hit areas, floods make an aquatic wonderland of terrestrial vegetation: fish swim among spectral tree trunks in the turbid water. The crowns break the waterline, placing the canopy within reach of a boat, a perfect situation for arboreal biologists who want to avoid fumbling with climbing gear. When the floods retreat, naturalist Henry Walter Bates tells us, "the trunks and lower branches of trees are coated with dried slime, and disfigured by rounded masses of fresh-water sponges, whose long thorny spiculae and dingy colors give them the appearance of hedgehogs."

Soil structure and fertility can mean life and death to many trees. Borneo's most widespread soil type (a clay loam) supports numerous tree species. Sites with sandy soil support fewer tree species, though most of these species occur nowhere else. To the roving botanist, forests there represent striking ecological islands in an expanse of ever-varying rainforest trees, like reefs in a wide ocean.

Some monodominant forests occur in habitats so locally specialized that most competitors are excluded. Areas subject to saltwater flooding may contain one or a few species of mangrove trees—perhaps the only tree species capable of surviving there. Yet adversity to invading species, if present, is not always obvious. *Gilbertiodendron* forests in the Congo seem to occur on the same range of soil types and slopes as the diverse forest tracts adjacent to them.

Monodominance may often have as much to do with biological as physical factors. In less windblown parts of equatorial Asia, for example, some of the most fertile soils harbor a surprisingly poor tree diversity, often one or a few dipterocarp species. It so happens that these species pack their crowns tightly together, throwing the understory into gloom. Saplings of most species starve from a dearth of light; only the monarchs of the canopy propagate successfully. This can lead to a simple, fixed community. Tree species that succeed on poorer Asian soils happen to produce an irregular canopy, thereby making room for a great variety of saplings.

Inch by inch, yard by yard, subtle physical changes in moisture, substrate, and slope can determine the odds for the survival of a particular species, even within a given type of rainforest. But, overall, it seems that the broad features of the physical landscape sustain a particular type of tree community in an area.

Biological Environment

In species-rich forests, how do so many rare trees coexist? Rare trees may simply be specialists in physical habitats that happen to be rare. Yet with as many as a thousand tree species in one small area of species-rich forest, it seems implausible that most could have unique physical requirements that are fulfilled only at small scattered sites.

The classic view of species competition is that tree species with similar requirements (rainfall, soils, and so on) shouldn't intermix for long. Regardless of whether seeds are tossed over the landscape by wind or hungry animals, most come to earth near their parent; ecologists call this clumping a "seed shadow." Most seeds therefore germinate close to others of the same species. If a tree species is the superior competitor for those physical conditions (sequestering the most water, light, and nutrients), it should come to monopolize the site, crowding out all others. Yet stands of one species—so commonplace in the temperate zones, where there are relatively few species in the first place—are the exception in rainforests.

Some scientists propose that tree species may compete equally, or that the competitive favorite at a site changes over a tree's lifetime due to environmental changes. Either way, no species is predestined to win out in a given location. Instead, the relative proportions of competing species would fluctuate unpredictably.

But in neither of these scenarios is species diversity likely to be maintained over the long term. In time, forests would lose most of their diversity as tree species went extinct because of competitive inferiorities or even chance.

Perhaps some of the trees at a site have bypassed competition entirely by taking root in habitats that barely meet their requirements for survival. Not ordinarily contenders for that habitat, they stay alive by luck. The seeds from which they sprang may have come from adjacent forests where other conditions prevail and those trees thrive. Such chance invasions presumably bump up species numbers, but scientists don't know how often species truly fit into this category. After all, any seed that makes it to the adult stage does so by luck—a centuries-old tree has to produce only one such seed in its lifetime to replace itself successfully. In a few years of human study, it's hard to prove how successful at reproduction trees in apparently "marginal" habitats will turn out to be.

Can many kinds of trees succeed in one area if they are not competitive favorites there? Possibly, with unwitting help from their competitors' enemies. Many plant-attacking insects and pathogens in the tropics favor one plant species or a group of related species, having adapted to the chemical defenses of those species. Buildup of specialized pests around a tree could make it hard for offspring of the same species (or group of species) to survive nearby, even if a site is ideal. This may give ecologically similar species that are immune to the pests a chance to establish themselves there, despite their competitive inferiority. The same thing applies to trees in temperate regions, though survival there may be dictated more by severity of physical (seasonal) conditions than by species interactions.

After maturing to full size, the single-seeded fruits of a *Platypodium elegans* tree will still require two or three months to dry before breaking from the branches and spinning to the ground. *(Photo: Darlyne Murawski)*

Some observations of pests on seeds and saplings support this hypothesis. Consider the *Platypodium elegans* trees of Barro Colorado Island (BCI), a reserve in Panama's canal zone run by the Smithsonian Tropical Research Institute (STRI). The trees are rare but their trunks, with innumerable convolutions, are unmistakable. In the dry season *Platypodium* trees drop leaves and unfurl yellow flowers to which canopy bees swarm. Seeds turn from green to beige in the year it takes them to ripen, then helicopter away on a sail-like wing. *Platypodium* crowns are so lofty (150 feet up) that prevailing northeast winds drift some seeds through the canopy ocean until they land 300 feet off, yielding an elongate seed shadow on the ground by each tree.

At close to four inches in length, the four or five thousand seeds from a *Platypodium* tree are conspicuous on the earth. They cram the litter beneath their parent; very few lie farther out. Yet seeds near the parent tend to be infested by gray weevils and blotched with fungi. Later you won't find one *Platypodium* seedling near the tree—only competing species survive there. The *Platypodium* sprout in isolation one hundred feet or more away, unmolested by pests.

If pests help maintain species-rich tree communities, this pattern of survivorship would have to be the norm for the trees that live in them. One testing ground for this idea is a 120-acre plot on BCI. I visited the plot with Robin Foster of STRI and Chicago's Field Museum, who developed it with Stephen Hubbell of Princeton. Robin's knowledge of trees is legendary, putting him in constant demand by students faced with a confusion of flora. His team on BCI has mapped every plant one centimeter or more in diameter at chest height. Dog tags gleamed on each stem or trunk, even where undergrowth appeared impenetrable.

A census has been taken of the plot every five years since 1980. "We miss a few plants, things that grow up and die in that period," Robin says. "But we are following the lives of a quarter million plants." The third census proceeded during my visit; it required a year of full-time labor by a dozen people. The data accumulated in the first two censuses compare in information "bits" with those contained in my twelve-pound Webster's unabridged dictionary, and this just to record each plant's location, identity, and girth.

Research on the BCI plot has generated an unparalleled range of scientific publications. Those regarding tree dispersal are consistent for the most common species: saplings do tend to root away from adults of their species. The effect is less noticeable with scarce trees, for most juveniles grow near their mother compared with distances between adults. When we take a closer look, however, seeds and

Trattinickia seedlings in Panama, like tree seedlings of all rainforest species, will contend with disease, herbivores, competing trees, and changing light conditions before reaching adulthood. *(Photo: Darlyne Murawski)*

seedlings near the adults can still be wiped out, perhaps by pests, and in time it's possible that only the most isolated individuals remain. This mortality would keep the trees from forming stands, and so help assure the coexistence of many trees within the forest.

This isn't scientific esoterica. The rubber tree comes from the Amazon basin, yet the region generates little rubber because of a native fungus blight that creeps over leaves and keeps trees from growing close together; rubber must be tapped from naturally occurring isolated trees in the rainforest. No such disease exists in the Old World, permitting the cultivation of vast single-species stands of imported rubber trees. The result has been an egregious financial loss for Latin America.

So how do natural monodominant forests persist? These often contain individuals of all sizes. The trees, then, are propagating, suggesting that the species will continue to dominate. Offspring in a stand of their own species must on occasion avoid the diseases or insect pests that infest their neighbors. Monodominant *Gilbertiodendron* and dipterocarp forests produce such bumper crops ("masts") of seeds and saplings that pests simply fail to destroy them all.

Moreover, microscopic fungi called ectomycorrhizae envelop the roots of many monodominant tropical trees. They keep the roots healthy, provide them with nutrients, and may speed intake of water. This gives such trees a competitive edge over those with less effective forms of tropical root fungi. It may be that saplings must sprout in a patch of their own species to obtain the fungus. This advantage may outweigh any increased disease or insect damage they suffer.

Treefall Gaps

Affecting both the physical and biological environment, the death of adult trees is as important in shaping rainforest tree diversity around the world as the mortality of seeds and saplings.

The gunshot sound of splitting timber, the creak of cable-thick vines stretched to the limit, the rumble and boom as a tree cleaves the canopy and strikes earth, the ensuing downpour of canopy plants, animals, and debris—all are unmistakable. Newcomers to the rainforest assume they've heard something special (after all, the tree must have taken centuries to grow!). In truth, given that several

FACING PAGE: Light pours through a canopy gap in peninsular Malaysia. Gaps come in all shapes and sizes and need not even extend to the ground; each offers unique opportunities for the upward growth of some plants and the lateral growth of others.

Looking like oceanic ripples frozen in time, the irregular surface of the New Guinea cloud forest probably reflects erosion of the underlying bedrock. Differences in drainage with slope are likely to influence the composition of tree species in a forest.

hundred thousand adult trees root within a couple of miles, treefalls within earshot are routine; some people boast of near misses. Biologist Don Perry wrote of a stormy night in the treetops in his 1986 book *Life above the Jungle Floor:*

> Suddenly a bloodcurdling groan overrode the shrieking winds. It was the sound of wood stressed to the limit—a tearing, snapping trunk. The tree was falling, and in a panic I untied my safety line with thoughts of jumping to another crown. The sound grew louder, almost deafening, like a freight train crashing through wood-frame houses. This was followed by a violent jolt; a giant tree had crashed to the earth and splintered into fragments. A flash of lightning froze the forest's new profile. There was now an immense hole directly in front of my platform where a [neighboring tree] had stood.

Toppling trees rip chasms in the sea of leaves. The patchwork of light created by these chasms is as significant to tree survival as the patchwork of soil types on the ground. If, after a canopy opening is created, conditions happen to remain suitable for the species that stood there before—by providing, for instance, the right levels of light and humidity for its seedlings—this species or one of its competitors has a chance to take over. But if the environment has become inappropriate to those trees, an ecologically different species will move in. The causes of tree death—whether they be wind, floods, or fire—and the qualities of the canopy gap that each loss creates are major thrusts of research today.

Ecologists have found that many deaths result not from old age but from disturbances to the forest. Storms are a commonplace example. Water collecting on a tree's leaves and bark and in the mats of arboreal plants may add literally a ton to the weight of an unbalanced tree; winds twist and bend its trunk and branches; a lightning bolt may deliver the coup de grace.

While hiking in the plot with Robin Foster, we entered a steamy, hot clearing formed by a recent treefall. A 140-foot kapok, beloved by BCI residents, had snapped midway up the trunk in a fierce storm. The broken surfaces were jumbles of twisted splinters the size of planks. The tree had smashed several hundred plants that Robin's team had painstakingly tagged in an area the size of two tennis courts. Raw sunlight streamed through the torn canopy and emblazoned the fallen wreckage like a floodlight.

Spacious treefall gaps, caused by the collapse of a huge tree like the kapok or the domino fall of many small ones, alter the local environment, providing an opportunity for invasion of "pioneer" trees (or "gap specialists," species needing direct sunlight to sprout). Pioneers become stunted and die if they are completely roofed over, but in gaps they grow fast, die young, and reproduce vigorously. Their tiny seeds scatter widely, then lie dormant until a large gap lets in light.

When I returned to BCI ten months after the kapok fell, an extravagance of thin-stemmed pioneers had grown up into the gap and stood taller than me, hiding the shattered remnants of the kapok. These intermingled with grasses and herbs, more conspicuous here than in the forest proper. Species different (perhaps more sensitive to heat) from those at the sunbaked center of the gap had sprouted at the shadier margins.

The old kapok was of a pioneer species that is exceptionally large and long-lived. Kapoks live so long that in time they loom over the majority of trees that form the mature canopy. (Regardless of whether or not such an oversized tree is a pioneer species, ecologists call it an emergent.)

I saw kapok saplings jostling for space around the sundered adult; one of them may replace it some day. This would be atypical for a pioneer species. Normally, by the time pioneer trees reach adulthood and die, shade-tolerant saplings (climax or "old growth forest" species) have grown up through the dim understory, where they are in a position to replace the pioneers. This sequence of events is called succession. Because the gap this old kapok produced is large enough to support the germination and rapid growth of its own kind, the young kapoks will be able to circumvent succession.

Climax tree species tend to live longer than most pioneer trees. Their seeds are often large, and so, as with *Platypodium*, land relatively near the parent tree. Instead of staying dormant in the understory, the seeds sprout and the saplings—sustained initially by stocks of food in the seeds—may wait decades for a chance to sprint to the canopy. To accomplish this, they may require just a glimmer of sky opened up by the fall of a small tree or branch.

Until there is a more severe disturbance, climax species continue to reproduce at a site, generation after generation. Yet when major treefall gaps occur, lifetimes of pioneer trees must pass before the return of climax species.

Pioneer and climax reproductive strategies are extremes in a continuum of tree lifestyles. While trees in temperate zones can also be categorized this way, early succession in the tropics involves far more species, some of which specialize in intermediate stages.

The size of treefall gaps may help account for differences in rainforests from place to place in ways we've hardly begun to understand. Many Latin American trees die violently, either by being uprooted or snapped in half like the kapok. For reasons unknown, moribund Asian trees tend to stay upright, gradually disintegrating in place. But even the overturned ones seldom kill neighbors: few

lianas link them, and with their compact crowns, they slice cleanly like knives through the forest. Thus Asian treefall gaps tend to be small, confining pioneers to small spaces.

Adult pioneer trees testify to past disturbances to the forest. If I were to return to the kapok site again in forty years, I wouldn't find a trace of the fallen tree. The forest there might not look much different at first glance from that around it, but I could tell where the canopy opening had been by identifying the trees.

Pockets of pioneer trees often comprise 3 to 5 percent of rainforest land; Robin Foster believes the figure could be even higher. Most rainforests are dappled with them, each one a patch of vernal greens that signify rapid growth of the flora. A randomly chosen spot in a rainforest mosaic is likely to have experienced a gap sometime in the last century or two, though some individual trees scattered through the forest may be centuries old.

Monodominant trees often persist in benign areas—places where severe disturbances (such as fierce storms that kill many trees) happen to be extremely rare, so that pioneers and other competitors have little opportunity to move in. Such trees are climax species that reproduce well in their own shade. Many ecologists argue that environmental conditions are seldom so stable, which may be why these trees—with their effective fungal associations and bumper crops of seeds—haven't taken over the planet.

In fact, in most forests environmental disturbances continuously make way for new species. If disturbances are too common or severe, however, diversity isn't necessarily high. Indeed, the word *jungle* refers to degraded rainforests of a type common near villages or logging sites. In a jungle the forest's cathedral effect, depicted so floridly by many authors, is lost. Instead of a dim expanse cut

Disturbances, Diversity, and Jungles

Early ecologists looked on treefall gaps as if they were a blight on the skin of a healthy forest, in need of healing. But we now understand that, when present in moderation, gaps help maintain diversity by fostering a patchwork of dissimilar species.

Only two months after typhoon Omar slammed into Guam's ocean-front forest in 1992, new trees have begun to replace fallen ones. Yet many of the canopy trees in storm-prone areas are adapted to resist high winds, and so quickly resprout their lost foliage.

FACING PAGE: With a fisheye lens, José Luis Machado photographs the canopy near a giant *Anacardium excelsum* tree on Barro Colorado Island, Panama, in order to measure the light fleck pattern above that spot. This pattern determines the light levels experienced over the day by the plants there, such as the *Anacardium* seedling next to José Luis.

Like many pioneer trees, *Cecropia* from Costa Rica has a simple branching architecture. Fast-growing pioneers often produce cheap, flimsy wood. A branch of ironwood (a climax tree) sinks in water; that of balsa (a pioneer tree) is famously light. *Cecropia* takes another tack: its flimsy trunk is hollow, providing living space for ants that protect the tree (see Chapter 6).

by shafts of searing sunlight, fast-growing herbs and vines adapted to high light situations choke the ground, while pioneer trees rule the canopy. Natural jungles form in the aftermath of frequent earthquakes, hurricanes, or other events that destroy whole forest tracts, or at wind-blown sites like forests bordering a river.

A jungle may also consist of a monodominant stand of trees, as when one pioneer species takes over near a village. In contrast to monodominant climax trees (like *Gilbertiodendron* and the other examples mentioned earlier), stands of pioneers—such as New World balsa trees—are replaced by climax trees if high levels of disturbances cease. Even so, in the few cases where pioneers are long-lived, one generation of them may last for centuries.

In short, except under extreme physical conditions, the occurrence of few tree species in tropical rainforests indicates a site that's either very stable or very unstable. In either case, the rules of competition have been fixed long enough for a few superior species to have won out, eventually excluding all others.

Most rainforests undergo intermediate levels of disturbance. Under such conditions, the rules of the survival game change unpredictably: the best competitor for any one spot depends not only on such things as soils, pests, and root fungi, but on the environmental history of that spot. There can be no winner for the forest as a whole. As a result, the trees of both pristine and disturbed habitats flourish together, and the total number of tree species and other organisms reaches its zenith.

Our perception of rainforests as forests primeval has been shaped by the awesome physical dimensions of many tropical trees and the meagerness of our lifespans when compared with theirs. As the physician Sir Thomas Browne wrote in the seventeenth century—perhaps just one dipterocarp lifetime ago—"Generations pass while some tree stands, and old families last not three oaks." Such antiquity obscures our view of the forests, and especially of imposing equatorial forests. As a few doggedly persistent field biologists have shown, rainforest structure and diversity is the result of variability and disturbance, not constancy and tranquillity.

FACING PAGE: In the koa forests in Kauai, Hawaii, native *Acacia koa* trees dominate. Tropical islands are often species-poor as a result of geographic isolation.

CHAPTER THREE

A PALACE OF MANY FLOORS

An arrow sliced upward through the crisp forest air. From my vantage point on the ground, I lost sight of it in the foliage almost at once. But the fishing line that trailed it unraveled from around a glass bottle at my feet with a steady, reassuring whisper. A heartbeat later I heard a whop! *as the rubber-tipped arrow ricocheted off an unseen tier of vegetation. The fishing line's whisper faltered ominously and died. Standing next to me, Pierre Berner lowered the bow and tested the limp line: "Almost!" He bemoaned the difficulty of this initial step in rigging a climbing rope in a tree.*

At 9,000 feet, sheer tree-clad slopes of Costa Rica's Talamanca Range loomed over us. A few plant species dominated the landscape: bamboos surrounded us like great quilled pens, while tropical oaks soared above. Pierre studies forest ecology here, in the upper watershed of the Rio Macho.

The equable conditions, with fertile, well-drained volcanic soils and an absence of storms and high winds, may explain why trees grow taller here than in most other cloud forests. The canopy reaches a 150-foot height, comparable to that of many lowland rainforests. Pierre wants to understand how ground slope influences the growth and design of these brashly towering oaks.

Some buttressed tree species have no central bole at ground level. The buttresses form early in life and enlarge and elaborate without further growth at the core.

FACING PAGE: An uncluttered view of the rainforest superstructure occurs where the riverbank has fallen away along a tributary of the Amazon in Peru.

PAGE 46: Pierre Berner paints rings around oak tree branches that hang into a chasm in the Talamanca Mountains of Costa Rica. He monitors the tree's growth by periodically checking the diameter of the branch at the rings.

PAGE 47: There are about 700 species of tree ferns, many of which are pioneer trees. Overlapping fronds form their spiral crowns (Papua New Guinea).

Trained in natural forest management in Switzerland, Pierre left his homeland in 1973 to work for development agencies in the American tropics. At that time tropical rainforest ecology was in its infancy. He found his education wanting. "I think the normal forests are in the tropics. The crazy stuff happens in temperate zones, with their seasons," he told me, echoing an opinion common among biologists. "But most of the knowledge we had then was on Europe and North America. It was conceptually good but not applicable. I needed to know more." He entered the doctoral program at the

University of Florida to study tree design. Since then he has come to know the oak-bamboo forests of Talamanca very well.

Like other rope climbers, Pierre shoots a fishing line over a branch as the first step in climbing a tree. Normally, with a compound bow, crossbow, or slingshot, this is straightforward—when you are lucky the first time. This wasn't one of those days.

On the next shot, the line caught in the abundant canopy plants on the branch. The arrow dangled beyond reach, and we were forced, after lots of frantic pulling, to cut the line.

On the third try, the arrow arched over the desired branch and plummeted from sight. Following cheers for Pierre's marksmanship, we faced the frustrating task of finding the arrow on a steep incline: a needle in a knee-bruising haystack. We searched for an hour before discovering that it had dropped not only the 120 feet from branch to tree base, but 200 feet farther downhill where it stuck in scrubby growth. Against the hazy sky, the twenty-five-pound-test fishing line was a gossamer thread stretching to Pierre's chosen branch. Aided by three friends, we struggled over rocky slopes to maneuver long lines and raise the climbing rope, something Pierre has done alone hundreds of times.

First we tied a parachute cord to the fishing line and pulled that over the branch, reeling in the line for use on other trees. Unlike fishing line, parachute cord is strong enough to bear the weight of a climbing rope. So Pierre tied the rope to the cord and the rest of us hoisted it over. "Careful!" Pierre cried from far uphill as he guided the rope's ascent: the strain of our effort combined with the rope's weight might break the cord. We tied the rope securely to the base of a tree. At last all was ready for Pierre to climb into the canopy the next morning.

I returned with Pierre and friends to Pierre's

cabin, custom-built for him near the forest. It was distinguished by a connect-the-dots pattern of tiny, oddly shaped windows, many of which had hinges. Admiring the eccentric handiwork, we warmed ourselves by a fire. Pierre had not yet installed plumbing; in any case we were too frozen to brave buckets of icy water. After three days our hair had turned to straw. The accumulated grime on our skin made us look as if we had tans. But given our current situation, appearance seemed trivial.

Later, at the home of Pierre's knowledgeable field hand, Elieser Mena, we ate *gallo pinto*—classic Costa Rican beans and rice with fried bananas, chicken, and shredded cabbage. The meal tasted great, though we reminisced about steamy baths and crepes suzette in Costa Rica's capital, San José. We discarded the prosaic discourse common among scientists in the lab and office for a more freewheeling conversation. Talk of tree biomechanics alternated with speculations on such matters as the best method for hoisting a cow into a tree. A canopy cow, Pierre decided, was just the photograph I needed.

The next morning, hands cramped with cold, Pierre pulled on his harness, from which dangled straps and two metal "ascenders." Then he snapped the ascenders onto the rope, one above the other, and began his ascent. When he had gone about a yard, two of us grabbed his harness and bounced him up and down—pulling the rope taut and testing the branch far overhead. The treetop shook vigorously, but the branch held: it would support his weight.

Several minutes of climbing later, Pierre dangled hundreds of feet above a dizzying abyss. He steadied himself with a grip on the branch overhead, cheerfully shouted, "The view is terrific!" and dislodged cherry-red bromeliad plants with sweeps of his hand. As he checked the branch's health and suitability for experiments, we shouted instructions back and forth until my voice grew so hoarse I could barely speak in the week to come.

Up went a bag of supplies; up went a line holding a can and a brush. Pierre, strapped in snugly, pried open the can, dipped the brush, and dabbed paint at intervals around the branch. Bands of paint allow him to plot growth by measuring circumferences at precisely the same spots on the branch at each visit.

To us terrestrians, Pierre's minute form, chiseled clearly by limpid air, with its helmet and brilliant orange and yellow safety colors, could have been that of a daft telephone repairman splotching mysterious symbols on a tree. Appearing ridiculous is a common fate for biologists in the field. Fieldwork surely tempers any arrogance bred by higher education.

Rock climbers devised the first ascenders. The method looks arduous, but with familiarity it becomes as simple as riding a bike. Starting at the base of the rope, I stand up in the leg slings (attached to the bottom ascender) and while bearing my weight with my legs, I use a hand to glide the top ascender as far as I can reach up the rope. The top ascender (attached to the harness around my waist) locks onto the rope at the new, higher level. Now I rely on the harness and let the top ascender bear my weight, so I can lift both legs into a squat again, raising the bottom ascender with my other hand as I go. By shifting my weight from leg slings to waist harness, from bottom ascender to top ascender, I can move quickly up the rope. It's an act of faith to trust the ascenders, but when I put my weight on them, they grip the rope tightly enough to hold me indefinitely.

I recall one of my first days of climbing when, at the most nerve-racking height, I happened to glance at the cord connecting my top ascender to my harness. This cord supports a climber's weight. As I watched, the knot in the cord pulled apart. I felt faint as it popped open. Only then did I understand nothing was awry—another knot actually bound the parts together.

Trunks and Foliage

Banks of foliage disguise the structure of the canopy from observers on the ground. For people like Pierre, who rise above the forest floor, differences in structure are obvious and hold clues to countless puzzles. Differences occur at two levels: the foliage and branching pattern of individual trees; and the vertical and horizontal spacing between trees. The botanist Edred Corner argued in his 1967 article "On Thinking Big" that:

> The world-stage of plant evolution has been the forest. We want to know about trees, their physiological feats, their strains and stresses, their reactions to all the animals that use them, their flowering and fruiting. . . . Botany needs help from the tropics; its big plants engender big thinking.

The life rhythm of each and every canopy organism is affected in some way by tree design. In Italo Calvino's *The Baron in the Trees*, the rebellious nobleman Cosimo discovers this when he forsakes the ground for an arboreal existence. He experiences profound differences among the structures of various tree types:

> The olives, because of their tortuous shapes, were comfortable and easy passages for Cosimo, patient trees with rough, friendly bark on which he could pass or pause, in spite of the scarcity of thick branches and the monotony of movement which resulted from their shapes. On a fig tree, though, as long as he saw to it that a branch could bear his weight, he could move about forever. . . . The fig tree seemed to absorb him, permeate him with its gummy texture and the buzz of hornets; after a little Cosimo would begin to feel he was becoming a fig himself, and move away, uneasily. . . . Sometimes seeing my brother lose himself in the endless spread of an old nut tree, like some palace of many floors and innumerable rooms, I found a longing coming over me to imitate him and go and live up there too; such is the strength and certainty that this tree had in being a tree, its determination to be hard and heavy expressed even in its leaves.

The canopy voyager feels these differences, but with an effect multiplied manyfold. Rainforest trees can be audacious in their height, bulk, and strength (many could indeed hold a small palace), or they can be spindly and weak. The number of species can be phenomenal, or one may dominate. Vines and other canopy plants can be scarce, or their omnipresence can add to the visual feast. This astonishing variety transforms all tropical tree climbers into epicures of canopy life.

For the land-bound observer, structural supports like buttresses and stilt roots are the most remarkable feature of tropical trees. These vary from wall-like flanges as much as three stories tall to woody networks connecting the trunk with the ground. They add immeasurably to the majesty of the forests. Among their functions, buttresses give trees a secure footing even on rock surfaces where the soil coat is thin. Ironically, their value in keeping tropical trees upright becomes most obvious after a tree falls, throwing upward a plate of roots several yards wide. We see then that most roots hug the surface soil; there are few deep roots.

Tropical trees often have smooth bark, but appearances vary remarkably. As ornithologist Frank Chapman wrote regarding Panamanian trees, some boles "are ponderously columnar," while others (perhaps *Platypodium* or *Alseis*) "suggest stripped athletes, with muscles and sinews swelling beneath their thin skins." In 1950 geneticist Theodosius Dobzhansky commented with equal relish on the trunks he saw along Brazil's Rio Negro and the Amazon:

The variety of lines and forms in tropical forests surely exceeds what all surrealists together have been able to dream of, and many of these lines and forms are endowed with dynamism and with biological meaningfulness that are lacking, so far as I am able to perceive, in the creations exhibited in museums of modern art.

A few trees have trunks with spines or other deterrents. Consider the gympie tree, common in Australia's Cape York. When I visited that area, I would casually mention the gympie to the odd person—waiter, cashier, gas station attendant—and was always rewarded by a memorable response. Opinions about trees are generally positive or at worst neutral, but gympies are absolutely despised. In hushed tones Australians tell of a friend who sat on a gympie log, or one who grabbed a gympie trunk for support on a hillside. Toxins, exuded by hairs all over the tree, inflame human skin. The pain may not let up for months. Gympies are not among those attractions described in tourist brochures. I've seen foreigners about to enter their first tropical Australian forest looking dumbfounded by placards that read, with no further explanation, "Beware of Stinging Trees."

Scanning up the trunks of tropical trees and into their foliage, we find that in climax forest the ubiquitous tree leaf (in some plants more accurately called a "leaflet") is of middling size, smooth, oval, and slightly leathery (but not so leathery as the leaves of many canopy plants), with a pointed tip that seems to aid in draining water. Whereas the fancy toothed or lobed leaves of the temperate zones are

uncommon, novel trees like palms, tree ferns, and bamboos greatly enhance the rainforest tableau, as do the many pioneer trees with oddly shaped leaves.

Though isolated trees may change color, summer greens dominate the palette. The lovely seasonal shift in tints that temperate-zone residents appreciate is absent from tropical rainforests. As with white in the Arctic, green is so pervasive that it may gradually overwhelm. "I can't remember the grays and browns of the temperate regions," biologist Charles Hogue recorded while in Costa Rica's Osa Peninsula. "No, the green becomes ingrained. It seeps into your skin and saturates the retinas of your eyes. It is almost as if your whole visual system and brain were impregnated with the forest chlorophyll itself . . . and inside your head it takes over your thinking."

But while rainforests appear evergreen, we can still pick out bare tree skeletons. These are not necessarily dead. Some of the tallest (emergent) trees are truly deciduous, lacking leaves for weeks out of each year in association with periods of flowering and fruiting. More often, foliage disappears briefly. A completely fresh set unfolds within days before rivaling vines and trees below have a chance to fill the space occupied by the bare crown. Such trees are still classified as evergreen.

Other trees replace all the leaves within a portion of their crown at once, thereby keeping most of their crown green at any one time. Such clustered

The Smithsonian Tropical Research Institute operates a canopy crane in Panama. Scientists board the crane's gondola on the ground. With fluidity gained from long experience, crane operator José Herrera reels the gondola upward and steers it among the crowns.

"flushes" of fresh leaves are common in the understory. The new foliage is pale, translucent, or crimson. Without binoculars I easily mistake leaf flushes for patches of gaudy flowers. Alternatively, the leaves may be renewed continuously throughout a crown. Many upper-canopy species and most pioneers use this strategy, creating a truly evergreen effect.

Tree Spacing: Light and Design

Near Panama City, a red construction crane pokes incongruously through the variegated treetops in a forested park. Erected in 1989, the crane is the brainchild of Alan P. Smith of the Smithsonian Tropical Research Institute. Scientists gain access to all of the forest's volume (from ground to the uppermost twigs) within the 125-foot reach of the crane's arm. Alan himself uses it to investigate spatial patterns in the physiology of trees.

Alan and other botanists find that tropical trees vary profoundly in the details of their photosynthesis, namely in the way carbon dioxide consumption, oxygen release, and sugar production respond to light intensity. The structure and physiology of most leaves are suited to only a specific type of illumination. Direct sunlight may wilt understory trees, while shading by neighbors may kill upper-canopy species.

One might argue that all the diverse trunks, branches, and leaf swaths in a rainforest are designed first and foremost to permit trees, by hook or by crook, to reach up and grab light. Accomplishing this isn't simple. Beneath any sunlit leaf extends a cone of darkness: the leaf's shadow. The cones from a spray of leaves fade with vertical distance until the light is evenly diffused, as if it had passed through sunglasses. Leaves can extract only a fraction of the energy from raw sunlight, so even those leaves screened by others above may photosynthesize at full capacity. However, for maximum efficiency leaves must be widely spaced from those above. Such leaves are illuminated evenly, rather than splotched by shadows. When successive tiers of leaves dilute light, it is no longer productive for a tree to add further tiers to its foliage: understory trees often grow just one leaf layer at the tips of their branches. Pioneer species, drenched in sunlight, could produce multiple layers—though one-layered pioneers, by investing more in lengthening their trunk than in multiplying their foliage, sometimes outgrow these competitors.

Occasional lacunae between branches let in shafts of light, casting motes of illumination on lower tiers. While these sun-flecks are exasperating to humans trying to take evenly exposed photographs, they provide vital energy sources to understory plants.

At ground level, any remaining ambient light has percolated through or bounced off layer upon layer of red-and-blue-absorbing foliage until it appears tinged with green. Less than one percent of the light intensity at the treetops may be left, amounting roughly to the difference between harsh midday sun and the illumination from a double-D-cell flashlight aimed a few feet away. For an impression of the strength of such a flashlight, try using one under a midday sun. It is virtually impossible to detect the beam.

Pierre Berner's steep Costa Rican terrain offers benefits to plants gleaning light. Viewed from head-on, trees on slopes form shingles: each overlaps only the bottom of the one above. This exposes trees to more light than they would receive on flat ground, where only the trees' apexes are exposed to full sun. Rather than growing straight, boles on slopes tilt into the abyss and away from the slope, further augmenting the light to the crowns. Also, as Pierre's painted rings show, crowns become lopsided, as branches jutting toward the abyss grow swiftly into the light there, whereas those extending uphill remain stunted. On flats, the same species grow symmetrically.

The increased light available to lopsided trees on

sunny slopes means higher rates of photosynthesis—more sugar production. "The trees are greedy for sugar," Pierre says, "like people." Pierre finds that trees on slopes do indeed grow faster than those on flats, but the unbalanced nature of the growth puts them in a precarious position. Oaks counter the severe gravitational forces brought to bear on them by this asymmetry by enlarging buttresses on their uphill sides, which act like sustaining cables. Still, these trees topple more often than trees on flats. Consequently, sloped forests have more treefall gaps and are more dynamic than level forests. Tree structure, growth, and survival vary so much with the type of slope that Pierre thinks each patch of oaks has unique features: forest dynamics at Talamanca don't repeat from place to place.

Asymmetric growth also occurs in trees bordering treefall gaps. Small gaps may be filled by lateral growth of adult tree branches. In larger gaps this unbalances trees, often causing them to fall into the gap and so prolong the opening in the canopy.

The Architecture of Tree Crowns

In their highest boughs the world rustles," Hermann Hesse wrote of trees. "Their roots rest in infinity; but they do not lose themselves there, they struggle with all the force of their lives for one thing only: to fulfill themselves according to their own laws, to build up their own form, to represent themselves."

Building themselves up from water and thin air, trees represent themselves through a repetition of similar body units called modules, and thereby reach the light. In trees, modules are branches or well-defined sets of branches; in colonial animals rooted on the ocean floor—creatures with tongue-twisting names such as coelenterates, bryozoans, tunicates, pterobranchs, and entoprocts—modules form bejeweled chains. In both plants and animals the growth patterns that modules follow are simple and few but nonetheless can generate striking results.

On a larger scale, botanists recognize about two dozen tree-growth patterns, or architectural models, that are defined in terms of the branching format and the layout of the leaves and flowers. All the patterns are seen in the tropics. In fact, whereas any broad region within the temperate zones (like all of North America or all of Europe) contains only four or five architectures, a single tropical rainforest commonly yields four times as many. Trees from completely unrelated families often show the same architectural plan; even a foot-high herb can be no different in underlying design from a giant tree.

Some patterns are distinctive. Anyone can recognize a coconut tree's slender trunk surmounted by a single tuft of leaves. In another pattern, the limbs of tropical *Virola* trees radiate like bicycle spokes at intervals on the trunk. But most patterns are harder to envision. Try to describe the dendritic pattern of an elm or maple in midwinter; imagine how much more difficult it would be to ascertain the pattern of these same trees with foliage attached, as must be done with most tropical species.

Some architectures may relate to stages of forest succession, some may relate to whether a tree lives in the understory or the canopy; to an extent, then, certain tree blueprints may be adaptations to certain light conditions. Still, as yet we know nothing of the adaptive significance of most architectural types.

The basic plan of most trees is invariant from birth until death. Nevertheless, some trees change drastically due to the addition of modules, to shifts in proportions of various parts, or to modifications that result from flowering. "Even giants must begin as seedlings," reflected Malaysian ecologist Francis Ng, "but a mature tree of the upper canopy is not

PAGES 56–57: Like many other single-species stands, kapur trees in peninsular Malaysia neatly demonstrate crown shyness and suggest that the principle operates even for individual branches in the crown.

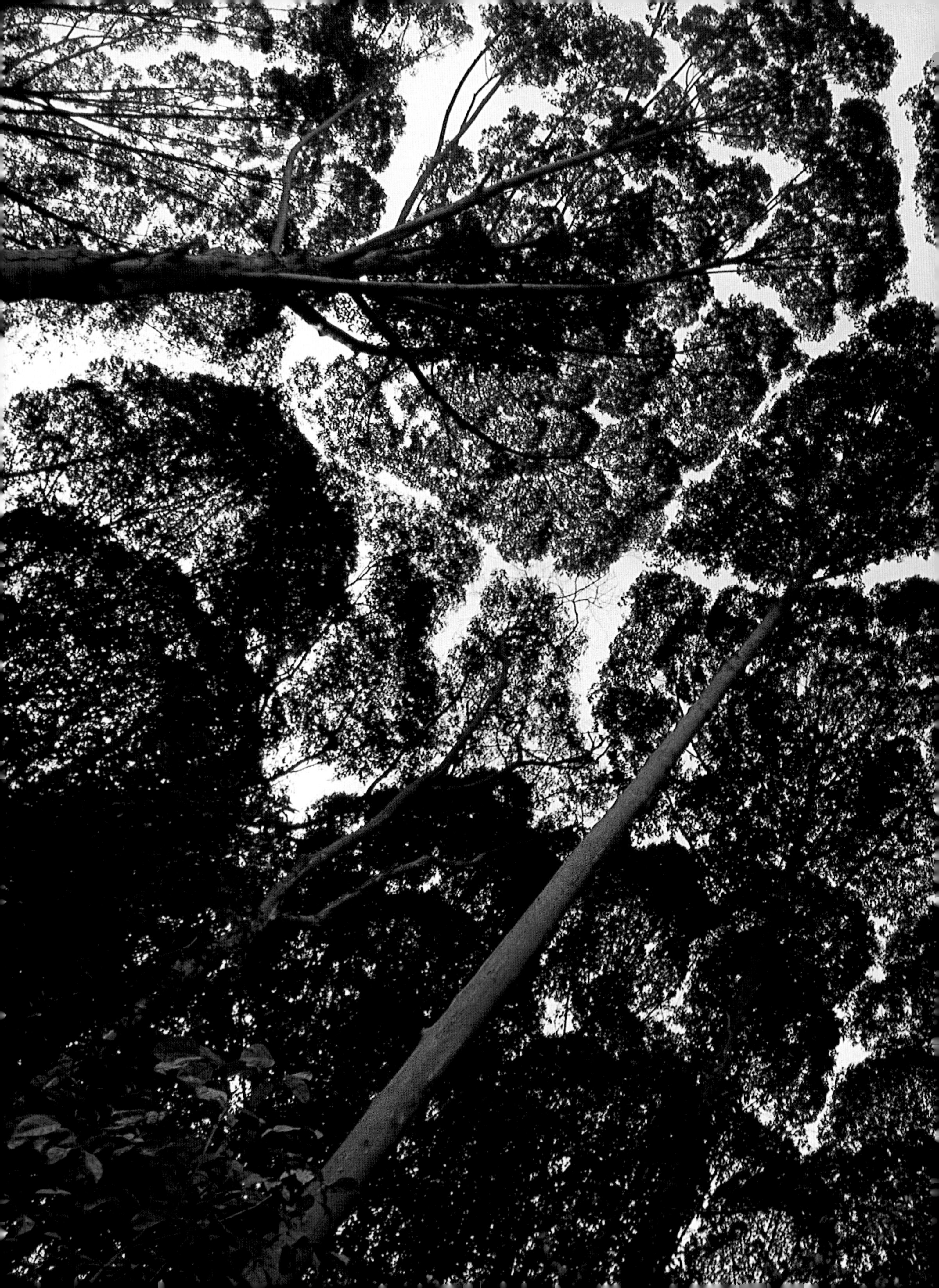

merely a seedling magnified a thousand times, any more than a skyscraper is a house magnified a thousand times."

Generally a tree adjusts to direct sun, high heat stress, and the magnitude of its increasing bulk as it grows out of the understory. In the shadows, its crown is deep and narrow with slender branches along the trunk; upon reaching its adult height, it transforms, directing its heavy upper limbs abruptly toward the sky and shedding the branches below. The result in a full-grown tree is a domed crown held aloft by vaulting branches at the apex of an unbranched bole. Such a mushroom shape is rare in the temperate zones, where trees in the upper canopy have rounded or (notably in the conifers of extreme latitudes) steep-sided crowns. These designs catch light from a sun that stays closer to the horizon than it does in the tropics.

In some tall tropical trees, architecture transforms as a tree thrusts its head from understory shade to full scorching sunlight. Leaves are no longer held flat out like solar panels to maximize light uptake; instead, they grow in whorls around the branches that let excess radiation pass through, reducing heat buildup.

A Sea of Treetops: Building a Canopy

The rainforest's many floors can be hard to perceive. Cruising a tropical river by boat, however, may be the best (and certainly most idyllic) way to encounter glorious views of stratification. Where currents undercut the banks, the tangle of riverside vegetation topples into the water, exposing fresh forest. In the mid-nineteenth century, Henry Walter Bates saw this process in action on the Amazon:

> One morning I was awoke before sunrise by an unusual sound resembling the roar of artillery. . . . The thundering peal rolled backwards and forwards, now seeming close at hand, now far off; the sudden crashes being often succeeded by a pause or a long-continued dull rumbling. . . . The day dawned after the uproar had lasted about an hour, and we then saw . . . large masses of forest, including trees of colossal size, probably 200 feet in height . . . rocking to and fro, and falling headlong one after another into the water. . . . It was a grand sight; each downfall created a cloud of spray; the concussion in one place causing other masses to give way a long distance from it. . . . When we glided out of sight, two hours after sunrise, the destruction was still going on.

A swath of fallen trees typically exposes a tidy cross section of the rainforest layers for our inspection. Stratification had been mentioned by explorers for some time, such as Georg Schweinfurth in his book *The Heart of Africa,* published in 1874:

> The gigantic measure of some of the trees was altogether surprising, but yet, on account of their various heights, their foliage lay as if it were in strata, and the denseness of the ramification wove the branches into a chaos as picturesque as it was inextricable.

Only in recent decades have methods been developed to plot forest stratification graphically. Most rainforest canopies are in such chaos, as Schweinfurth describes them, that some scientists argue today that true strata seldom if ever exist in the tropics.

Where stratification occurs, how does it come about? Just as leaves cast shadows that influence the location of the leaf sprays beneath them, trees influence the heights and positions of those below as each jostles to seize space. The pattern is dependent on the particular trees at each location and on their structure, their response to shade, the order in which they took root, and their response to treefall gaps and other disturbances.

Given this, the greater the diversity of tree species in the upper canopy, the less predictable the layering below them may be. Indeed, layering is simplest and easiest to discern in temperate zones, where often only two to four tree species dominate, forming a uniform canopy and a straightforward understory. In southern New England, for example, flowering dogwoods form an understory in forests of oak or other deciduous canopy species. Tropical rainforests dominated by one or a few tree species are similarly uniform.

Ecologists have distinguished as many as five strata in tropical rainforests with diverse tree communities, attributing this complexity to year-round growing seasons and to strong tropical sunshine. These conditions facilitate the growth of many tall trees, which in turn provides space for more layers beneath. The domed crown of the archetypical upper-canopy rainforest tree also allows more light to penetrate to strata below it than does the steeply inclined crown of a temperate tree. Because most tree species specialize on narrow conditions, the complexity of the layering clearly contributes to the local richness of tropical trees.

The five or so tropical strata are not cut-and-dried. Trees of all sizes exist; strata can be defined only insofar as certain size classes prevail. Most distinctive are the scattered emergent trees that tower over the rest of the forest, and the upper canopy beneath the emergents, where crowns lit by direct sun pack together. Most photosynthesis takes place here, in these two strata, the most productive part of the canopy. Emergents, facing uniform, exposed conditions, encompass far fewer species than upper-canopy trees, which face varying conditions depending upon the degree of shading from their neighbors. The kapok (*Ceiba pentandra*) is a common emergent in the New World and African tropics; dipterocarps of the genus *Shorea* are frequent emergents in Asia.

Below the sunlit layers lie understory trees, some adapted to shade throughout their lives, others juveniles of tall trees that jockey for space in the strata above them as gaps open. In either case, their crowns tend to have a distinctive deeper-than-broad shape. The density and heights of the sunlit layers determine the complexity of the strata in this dimly lit region, but the number of species involved is usually large. Below the understory lie sparse, loosely defined strata of shrubs (woody plants with multiple stems at soil level) and, finally, herbs. The surface area of leaves suspended above a square foot of ground from all the strata can be greater in a tropical rainforest than in the temperate zones because greater light permits more photosynthesis.

There may be some consistency to the layering pattern independent of the species present. A space exists between sunlit trees and those understory specialists directly below them for the same reason that leaves within a tree are spaced: trees growing too close to those above would be irregularly illuminated, with sections of their crowns always in shade. By ceasing growth before that height, understory species encounter a more uniform illumination across their crowns. The space above them is not entirely empty, however, because young climax trees destined for higher strata must continue to grow through less desirable, unevenly lit heights (often in bursts as gaps appear above). Also, because the species-rich rainforest roof is irregular, the understory layers will be irregular, too. Thus even where strata do occur, it can take an observer time and effort to sort them out.

Crown Shyness

In addition to its vertical stratification, the canopy varies laterally. For one thing, rainforest crowns seldom intermix; like a territorial animal, each tree defines its own area. Empty spaces may separate crowns, though detecting them can be difficult through the sundry overarching canopy layers. Ideal views of the spaces occur where young

trees have sprung up simultaneously to fill a canopy opening. All the largest trees are therefore about the same height. Under such conditions, some Asian canopies, with compact crowns and little clutter from vines, are particularly striking. From beneath they resemble a jigsaw puzzle of loosely fitted polygons, like the blotches on the flanks of a giraffe.

Scientists rather fancifully call this phenomenon "crown shyness," as if trees actively avoid each other, perhaps to slow the spread of pesky animals or to avoid bothersome, wind-induced encounters with neighbors. Certainly, existing spaces have such effects, but it's more likely that their formation is accidental. The tips of branches extending toward adjacent trees are often dead. Two theories offer explanation: dieback may occur where one limb shades another; or the wind, in rocking treetops together, may causing the wearing away of any outer branches close enough to touch. Both hypotheses may help explain why separations between tree crowns appear to be most distinct in the highest canopy tiers.

Researchers find that Costa Rican mangroves (a forest type characterized by tidal saltwater floods) have wider spaces between trees where the wind blows neighboring branches the farthest. This supports the wind-abrasion hypothesis. In fact, a range of factors may affect spacing. Vines, for example, might exert their own influence: coiled like springs to partially accommodate tree motion, they could moderate sway and so incidentally reduce gap size.

Less dramatic than a treefall gap and, as such, easier to ignore, crown shyness gaps also contribute in significant ways to forest ecology. In a stand of Malayan kapur trees (*Dryobalanops aromatica*), I cleared away the understory foliage to find a superb example of a shyness pattern. I estimated that the clear bands of sky between the trees made up from 7 to 13 percent of the canopy surface. As in most forests, little light filtered through the crowns themselves—perhaps one or two percent of full sunlight. At least here, most of the light available to the first layers of vegetation below entered through these narrow spaces between trees.

The gaps resulting from crown shyness seem two-dimensional to us when we can make them out at all from our secure and simple ground environment. But to canopy dwellers they present an imposing three-dimensional challenge: wherever trees approach one another, regardless of the angle or location, there tends to be an abyss. This means trouble for climbing plants and animals but provides a narrow entryway to strata above or below for airborne organisms.

Because few arboreal species stay in one tree throughout their lives, tree stratification and crown shyness profoundly influence how plants and animals survive and reproduce in the canopy.

Rising above a Frontier

I have come to keenly appreciate the structure of trees and canopies. Now when I climb a rope, I sense and respond to canopy strata. I feel nervous for the first few yards. The rainforest floor, still visible, gives evidence of my height—but at least it is familiar territory. Later, inchworming through the understory, I relax as the leaves seem to envelop me in a protective cocoon.

Rising past the foliage, I enter open air, with only the solemn trunks of upper-canopy trees around me. Height can't be gauged. Below I see only patches of near-black ground. I feel exposed and vulnerable. (I wonder, could open spaces between strata be as much an impediment to jittery climbing animals as a road is to many animals on terra firma?) Above me, the top stretch of rope has come into view, ridiculously long and fragile. Heart pounding, I imagine myself an astronaut in free-fall.

PAGES 60–61: Clutching twigs, Jack Longino pokes his head above the cloud-forest canopy—high atop a fig tree at Monteverde, Costa Rica—and surveys a tidy, wind-shorn topography of leaves.

Continuing my climb, I pass into the crowns of upper-canopy trees. Surrounded once more by foliage, I calm down. But this response to a rainforest layer bears no relation to actual danger: a fall here would be no less lethal than a fall from the open area below, or indeed from understory layers below that. My mind, however, finds comfort within cloaking vegetation.

In 1991, a year after I first climbed with Jack Longino, I returned to Costa Rica seeking spectacular views and novel canopy insects for the Museum of Comparative Zoology collections. Jack and I set off for an impressive fig tree that scientists there had named "Wilbur," which stands in the lower part of the Monteverde reserve. Although Wilbur's architecture consisted of comfortable, flat branches, the wind at this site blew fiercely. Ascending through the understory layers into Wilbur, I found myself gripping a dense carpet of vegetation as my branch both twisted back and forth and waved in circles.

Next to me Jack poked at the base of an orchid. "These gyrations are hard to take," he noted, with his usual talent for bland understatement. Suddenly focused, he drew an ant from the treetop soil with his forceps, examining it as one might a fine gem. He extracted an earthworm—how did that get up here?

Below us less mighty tree crowns surged within a receding sea of green: the wind was building. A turquoise hummingbird buzzed by. Soon all the world moved; everywhere branches groaned.

Most of Wilbur's leaves grew at the tips of the upper vault of branches, giving us a vista of the tree interior and the broad sweep of the understory. With our eyes adjusted to canopy glare, the receding forms of trees beneath us seemed saturated with a cavernous darkness. Far below, Wilbur's buttresses formed a dim star. Despite the profusion of canopy plants cloaking most of its limbs, we had an excellent view of the tree's branching pattern.

With typical nonchalance, Jack decided to reach a distant part of the fig's crown not by climbing down to another set of vaulting branches but by crossing the chasm separating the major divisions of the tree. For one tense moment his arms and legs stretched between two boughs that swung out of sync in the wind. When the wind died, he released his grip on the first branch and, with his safety rope arcing behind, stepped to the far side.

Later, vials of critters in hand, we studied the backlit limbs arching overhead that were buffeted by the wind. Wind is a problem for canopy trees, one not appreciated by pedestrians in the calm understory air. The forces involved can be awe-inspiring. (To reduce drag, the leaves of some trees alter shape during turbulence, forming cones; others collapse into streamlined arrangements resembling fish scales.) After some reflection on the power of tropical weather, we climbed until we had balanced ourselves near the branch tips, and thrust our heads past the uppermost foliage.

Before we could take in the view, the wind gusted. Our branch began to bend. Clambering down to sturdier boughs to gather our wits, we marveled at Wilbur's structure. Most temperate trees have outermost branches too slender and too intricately subdivided to support the weight of two full-grown men that high up. This tree, with its different branching pattern, was a tropical heavyweight.

The wind paused. Up we went into the twigs again, trying not to think of the drop below, twelve stories by urban standards.

Startling in the brilliant sun, a leafy surface receded over the curve of the fig's crown and continued from one tree to the next and on to the distant hills. It looked as if it had been manicured by a giant gardener: wind must shear any branches that poke above the rest. This, in the strict sense of the word, was the canopy. But to me, that day, it appeared more like a simple aerodynamic green carpet than the surface of the most structurally complex ecosystem to grace this planet.

CHAPTER FOUR

GARDENS IN THE SKY

As a diversion from two weeks of tracking monkeys in Uganda, I searched out, in a forgotten patch of forest near Kampala, an imposing climbing tower built in 1958, one of the earliest of its kind. No one seemed to have been there in years. The spiraling metal staircase through its center appeared secure despite rust, so I climbed 120 feet to a vista of forest and swamp. Hornbills argued noisily in the hazy orange light. I strolled the top platform, imagining the British researchers who once used the tower to investigate which canopy strata African mosquitoes selected for breeding and hunting blood. Making no pretense of technical sophistication, they used humans as mosquito bait.

Recently built towers into the canopy, while perhaps structurally more state-of-the-art than Uganda's dinosaur, can stand my hair on end. In 1990, Harvard student John Tobin and I met Jeff Luvall in Costa Rica as he prepared for work at one such modern tower. Jeff studies physical conditions in and above tropical canopies. Based at NASA's George C. Marshall Space Flight Center in Alabama, Jeff is fascinated by the effects of radiation, temperature, humidity, and other factors on plants. Pondering with us the mysteries of tropical atmospherics over beer, Jeff offered to show us his lowland rainforest tower in Braulio Carrillo National Park.

When we reached the tower two days later, it did not immediately inspire confidence. Barely twenty inches wide for its entire length, it resembled some sort of narrow radio antenna tower supported by guy wires. Jeff grabbed a rung and began his ascent. After an uneasy pause, John and I followed, climbing hand over hand.

Passing layers of foliage, I saw many kinds of plants that live only in the canopy. Among them are epiphytes, which root on bark or soil in the canopy, taking water and nutrients from wherever they can get them. Less abundant are parasites—mistletoes—which tap into the vascular system of the host tree for water and nutrients. Other species send roots into the earth just as trees do. In hemi-epiphytes, this connection exists only during part of the life cycle; they spend either the beginning or the end of their lives as epiphytes and are rooted to earth for the rest of the time. In climbers such as vines, the ground is tapped permanently. Together these plants form a community—until recently a virtually untouched garden of secrets.

After climbing steadily for several minutes, I poked my head above the treetops, 110 feet above the ground. The tower shot still higher. Forty feet overhead, Jeff uncovered a computer affixed to our spindly spire. A solar panel near the computer powered it. Above him, at the very top of the spire, a horizontal pole sported a wind vane, a spinning wind-speed gauge, and sensors I didn't recognize. Feeling like one ant passing another on a narrow twig, I squeezed by Jeff to take a closer look. From the pole the tallest tree crowns billowed fifty feet below. The ascent had been equal to scaling a sixteen-story building—without the building.

Jeff waved at the sky. "Storm coming!" He quickly covered the computer. The sky dimmed; moments later the scaffolding shuddered. Like an explosion, the rain had arrived with surprising speed and force. I could barely detect Jeff descending through sheets of water. Below him, John and the lower part of the tower were invisible in the driving rain. It seemed as if I were magically floating in space over a murky, seething jungle.

I snapped a lens cap on my drenched camera and lowered myself a rung. My foot slipped. On the way up my boots had easily gripped the thin rungs. A film of algae, then dry and undetectable, now made the metal as slick as ice. Rain formed walls around me; I imagined myself inside a waterfall. My descent would have to be unbearably slow. Rung by rung, I lowered my feet carefully, holding the tower with both arms in a vise-like grip.

I had almost reached the canopy several minutes

PAGE 64: Christopher Columbus believed that tropical American epiphytes were outlandish parts of the trees they occupied. While the leafy conglomeration of one tree does seem to form a complex individual, actually many plants are involved. Canopy growth in moist mountainous areas like Monteverde, Costa Rica, is more profuse than in lowland rainforests.

PAGE 65: In a Colombian cloud forest, a red *Guzmania sanguinolenta* bromeliad is flanked by strings of a radically different bromeliad species—*Tillandsia usneoides,* or Spanish moss. For water and nutrients, the first depends on rain flooding a central cistern; the second, on mist or rain droplets wetting microscopic scales on its surface. Most epiphytes, such as these two species, are found only in the canopy; their seeds perish if they lodge on the ground. The adult plants likewise expire if they loosen and tumble—as many eventually do—from their canopy roosts.

later when I became vaguely aware of something burning against my back. "Now what?" I thought, as I concentrated on each step. Soon it was intolerable; I felt behind me. Short-circuited by moisture, my fancy electronic camera was fiery hot. As I struggled to find a source for this heat, the camera began to make popping noises. That was all I needed. Delaying my descent, I took out its batteries with the anxiety of a detective disarming a bomb.

The canopy, like a leaky umbrella, reduced the rain's impact and shielded me from wind. I moved confidently now. I recalled Uganda's stable old tower with renewed respect as I reached terra firma, resolving to use a safety line if ever I did this again.

Later, warm and dry at a field station a four-hour drive away, I pondered the effects of rain on the tropical canopy. The first sound as a downpour approaches in a rainforest is the faint white noise of water washing over green expanses. The sky dulls. The hum of insects silences. Overhead, branches begin to clash in the wind. The tones of the rain grow ominous, until suddenly a roar arrives in waves from directly above. Darkness shrouds the foliage. But where is the water?

It takes a few minutes for the first drops to work through the vegetation. In rainstorms, I encourage companions to stop and observe their surroundings instead of sprinting to shelter. As the first droplets splash on the leaves above them, streams of water funnel along tree limbs and down the trunks, seeping out from the bases of the trees. Far overhead, some of this funneling water percolates through the fine root mats of thirsty epiphyte plants, which absorb much of it. During showers that would be heavy by temperate-zone standards, hardly any moisture may reach the ground.

We decided to try for the tower again at 2:30 A.M. to catch the sunrise. Groggily, we rose in darkness. Sliding along muddy slopes on an unlighted logging road, our Jeep felt like a toboggan, making the trip as stomach-churning as the climb down the tower had been in the rain. We could see only a fraction of the road at any one time in the narrow headlight beams. At the wheel, John did his best to predict curves and drop-offs before they loomed into view. The Jeep careened back and forth, barely under control despite the four-wheel drive and chains on the tires.

Our arrival at the base of the tower was timely. Up we clambered again, this time tied in with mountain-climbing ropes. As we neared the top of the tower, the sun cracked the green horizon like a slice of orange, exquisitely lighting an undulating canopy surface. I imagined the world awash in forest green as it had been centuries ago, before humans became greedy for that miracle material, wood. A few yards below Jeff made himself comfortable in a chair bolted to the tower. His fingers flew over the computer keyboard. Muttering technical terms, he copied data from the last forty-five days onto a cassette as the canopy rippled far beneath him.

A flowering *Stelis* orchid from Colombia roots in soil and depends heavily on mist for moisture. While canopy orchids like *Stelis* look like ordinary plants, other orchid species have lost their leaves and stems, being reduced to a green photosynthetic root from which pops the occasional flower. The stress such epiphytes face is easily appreciated by humans: a person spending a three-hour stint in the sunny canopy of lowland rainforest can easily drink two or three quarts of water, while a colleague working equally hard in the shady understory is still doing fine without a sip.

PHYSICAL CONDITIONS IN THE CANOPY

Studies by Jeff and others chronicle the physical conditions in and above the canopy. At sunset, the cooling air is able to hold less water vapor, and the excess moisture condenses into ethereal mists that rise from the trees. I experienced this myself in Africa during a night spent in the upper canopy. Soaked in condensation and chilled to the bone, I would have been more comfortable on the ground, which stays warmer because of the insulation of the canopy mantle. I awoke to the sound of rain, even though the sky above me was bright with stars. Curiously, the sound came from below: so much water had beaded up on the foliage that I heard the patter of what must have amounted to gallons of dew dripping from leaf to leaf.

Rain and dew would appear to provide more than enough moisture for plants, but the wealth of canopy plants belies the harshness of the environment. Drought stress on upper-canopy foliage during the day is acute: temperatures may soar to over ninety degrees Fahrenheit, while relative humidity may drop to below 60 percent. Such conditions may seem reasonable when compared with those of a desert, but in the intense equatorial radiation, plants rapidly lose water to evaporation.

Rainforest plants at ground level enjoy conditions as tranquil as those in a greenhouse. Because of the buffering effects of overtopping leaves and branches, humidity constantly hovers around 100 percent and temperatures are almost as uniform. But light is often wanting.

People, of course, rate rainforest comfort levels differently from the way plants do. I, for one, was never meant to live in a rainforest. The shady understory may be relatively cool, but perspiration in soupy air soon turns me into a wet dishrag (my body adjusts somewhat after a week or so). Canopy heat feels more pleasant, thanks to reduced humidity. But even in treetops I cannot win: tropical radiation turns me a startling shade of crimson.

The lifestyle of each rainforest plant—epiphyte, climbing plant, tree, parasite—has its drawbacks and advantages. For example, canopy sunshine promotes the survival of epiphyte seedlings. In contrast, the dearth of light energy near the forest floor often starves tree seedlings. And even when light from a canopy gap does sustain growth, the tree must expend most of its energy on the supportive tissue (wood) it needs to reach and take over that gap. Epiphytes, on the other hand, require little if any supportive tissue to help them get at light. Held aloft by trees, most epiphytes have the luxury of remaining small throughout their lives—they need not have legs long enough to reach the ground.

Nevertheless, epiphytes face other awesome challenges to survival. To counter evaporative loss, trees and climbing plants are able to tap into the moisture reserves of the forest floor, but the tiers of epiphytes must contend with wild fluctuations in their water supply. Epiphytes need to intercept water as it hits their tier, before it evaporates or drips beyond reach, leaving their branch bone-dry. Most temperate-zone plants would wither and die as epiphytes in the rainforest canopy—that is, if a tropical downpour did not smash them to pieces first.

SOIL AND ROOTS IN THE AIR

Nalini Nadkarni, a professor at Evergreen State College in Olympia, Washington, feels at home in the cloud-forest roof. On occasion, she carries her infant, Gus, into the canopy of Monteverde, Costa Rica, for a day's work. Today, however, Gus played in Nalini's cabin with a sitter while Mom swung about in the canopy.

As I looked on in the same tree crown, Nalini spread the greenery on a branch with her fingers, gleefully calling out multisyllabic Latin names for the epiphyte species before her. She counted and

low when I leaned back to rest. More plants dangled from branches above, filling my view.

The beauty of the site was not unusual for a tropical cloud forest, but the ardor of the scientists' investigations of epiphyte survival strategies riveted my attention on the plants. How can such an abundance of canopy plants obtain not merely enough water, but all the nutrients they need? The answer lies in part in the existence of humus in the canopy. But, curiously, the significance of this humus to the tropical ecosystem first became evident from studies of trees, not of epiphytes.

LEFT: Nalini Nadkarni, a professor at Evergreen State College in Olympia, Washington, studies roots and soil in the Monteverde canopy. Slicing through the bed of epiphytes on a branch with a pair of clippers, Nalini rolls back the plants and soil like a thick strip of turf. The hair-thin epiphyte roots hold the soil in a tight meshwork. She will monitor the reappearance of soil and plants on the stripped part of the branch.

BOTTOM: Tree roots extend through soil and moss in the canopy of a moss forest in Papua New Guinea.

measured each plant's leaves, which she had marked with dots of typewriter correction fluid two months prior. By this means, her research team obtains the first accurate assessments of the natural growth rates of these little-understood plants.

Teri Matelson, wedged snugly in a nearby fork of the tree, recorded the names and numbers in a well-worn notebook. In charge of Nalini's Costa Rican field projects, she will, over the years, revisit this site many times to check on how the plants have fared.

Certainly the environment appeared competitive for epiphytes: there was hardly a square inch of tree bark left bare. Plants, from algae splotches thinner than a fingernail to shrubs strong enough to support my weight, cascaded over the trees and atop one another. Epiphyte twigs jabbed my thighs; spongy plants oozed moisture through my clothing. Varied branches and leaves billowed up from either side of where I sat, making a scratchy but serviceable pil-

In 1980, Nalini became intrigued by trees that grew arboreal roots from branches in their crowns, fifty feet or more in the air. She noticed that the roots, which can be an inch wide, extended into humus impounded by the fine, intermeshing roots of a mat of epiphytes in the tree. Up to that time, no one had paid much attention to canopy humus, but Nalini found that nutrients from this aboveground source were being incorporated into the tissues of both the trees and the epiphytes. Not only that, but

the canopy roots developed some of the same symbiotic relationships with microorganisms (mycorrhizae or bacteria) that the trees used to extract nutrients with their roots in the ground. Canopy roots, in other words, were as functional as ground roots.

It became clear to Nalini that canopy humus formed a resource for the forest as a whole, a reservoir of nutrients that even some trees cannot afford to ignore, given the proverbial infertility of rainforest soils. Aerial tree roots have since been

documented in several tree species in temperate and tropical rainforests around the world. Yet most work on them, as on the canopy humus and their associated epiphyte communities, has centered on the New World tropics.

Where does canopy humus come from? Some epiphytes build up soil by entangling windblown litter in their roots. Others furnish living space for ants whose gathered food and nest construction materials eventually decompose. But these are minor pathways, Nalini suspects. She believes that, at least initially, most soil builds up in the canopy not from sloughing tree bark or from decaying leaves, but from residues deposited by the air.

In fact, some epiphytes don't require humus at all. Biologists near the turn of the century often called epiphytes "air plants"—an appropriate term for such species, for they cling to the bark surface, ostensibly surviving on air alone. Some hardy forms like certain bromeliads even survive on electrical wires.

Soaked to the bone by the tempest, Ken Clark, a University of Florida scientist who studies in Costa Rica's Monteverde Reserve, collects dew that has dripped along Teflon filaments into plastic jugs. The mist settles onto the cloud forests below him, where it adds substantially to canopy nutrients.

FACING PAGE, TOP: "Air plants" like this bromeliad on a bare branch depend heavily on mist deposits. CENTER: Teri Matelson investigates nutrient turnover in a canopy laboratory atop a fig tree in Monteverde. BOTTOM: Teri marks dead leaves with white spots and strews them over branches to determine the rate at which they fall out of the canopy. Almost no leaves stay in the treetops long enough to decay and add to canopy soil.

To an extent, "air plants" do indeed feed on the air, as I learned during my sojourn in Monteverde with Ken Clark. For his Ph.D. dissertation at the University of Florida, Ken was meticulously investigating how plants trap airborne nutrients. He used mist collectors: condensing droplets slid along foot-long filaments of chemically neutral Teflon into milk jugs. Helping Ken gather his jug samples in a pass between the Atlantic and Pacific sides of the mountains, I endured wind howling at sixty miles per hour and mist soaking my hair and clothes.

The mist holds nutrients crucial to plants (nitrates, phosphates, other trace elements, even organic matter). These are carried for a hundred miles from the Pacific Ocean before they reach Monteverde and wash over the canopy. Rain and dust can similarly hold substantial nutrients. Indeed, rather than depend on canopy humus, some trees and "air plants" have canopy roots that hang in the air. These absorb nutrients in the water dripping from leaves and soil above. "One way or other, arboreal plants snap up nutrients," Ken said, explaining that some waterborne nutrients arrive in the canopy at high concentrations, but by the time the water reaches the ground the concentration is markedly less.

Ken's study suggests that canopy humus takes decades to accrue in cloud forests, one particle at a time. Indeed, hardly any soil had accumulated where Nalini and Teri stripped the branches seven years before, and only traces of algae and moss had reestablished themselves. Seldom do lowland rainforests have canopy humus as deep as that in the much moister cloud forests, and, perhaps consequently, they have more modest epiphyte mats. Experiencing less mist, forests near sea level presumably accumulate soil more slowly.

After thriving epiphyte populations build up on humus, decay of their litter recycles nutrients in the canopy. Still, many epiphytes fall when branches break, or when mats of plants and soil slip from branches in miniature landslides. The nutrients these plants hold are lost to arboreal plants but are readily available to ground-rooted plants—including trees that convey the nutrients back to the canopy, incorporating them in leaves and limbs.

In fact, rainforest plants, in the canopy as on the ground, capture and assimilate nutrients so swiftly that, as writer Eric Hansen reports of Borneo, "it made me feel as if we were traveling through the intestinal flora of some giant leafy creature."

The Abundance of Epiphytes

"As a graduate student I first considered doing a lab study for my thesis," Nalini told me during a pause in her canopy inventory, glad to be in the field and not in a lab. "It wasn't me! I broke a lot of glassware. Then I had my first look at Monteverde in the summer of 1979. I hiked a trail called Sendero Bosque Nuboso, and the epiphytes overhead floored me. They were so beautiful and clearly so important to the forest, but no one knew much about them. I immediately wanted to climb." Laughing, she added that it took some doing to convince her academic supervisors at the University of Washington that treetop research for her dissertation was feasible; few people worked in canopies back then. Maybe she surprised her skeptical earthbound faculty with her ground-breaking thesis, which compared epiphyte communities in Monteverde with those of temperate rainforests in Washington State's Olympic Peninsula.

Though we may seldom think of canopy plants in the temperate zones, they are there. A few species flourish in temperate rainforests and swamps. Some even survive in deserts. But while plants in fertile ground soils can seemingly spring up overnight, epiphytes, using limited resources in the canopy, demand years of uninterrupted benign conditions to mature. Winter frosts make treetop existence problematic for epiphytes in northern latitudes.

In temperate forests, small plants burgeon not in the trees but near ground level. Because frost seldom penetrates the earth deeply, many small ground plants survive cold winters as underground tubers, corms, or bulbs. Enough light filters through temperate canopies to support a dense ground cover beneath. This is especially true in the spring, when herbs and understory plants grow vigorously before the tree buds above them burst open—an opportunity unavailable in tropical rainforests.

For her thesis, Nalini found that temperate and

tropical epiphyte communities differ strikingly. In the temperate zones and the elfin forests of the tropics, most epiphytes are lichens, liverworts, and mosses that can endure temperature extremes. In the tropical lowlands and cloud forests, these plants are also abundant in trees, but many more epiphytes are flowering plants. (One of the few flowering plants in temperate-zone canopies is Spanish moss, a stringy kind of bromeliad found in cypress swamps.)

A tenth of all "higher" plants (flowering plants and other species with vascular systems) have taken to epiphytic life in the tropics: more than twenty-eight thousand species in eighty-four families. In addition, an uncounted number of epiphytes are lichens, fungi, algae, and "lower" plants (those plant species without vascular systems, such as liverworts and mosses). Most epiphytic species exist only in the canopy.

Although epiphytes are a universal feature of tropical rainforests, they may reach a zenith in numbers and diversity on the slopes of the Andes. As part of the only complete botanical survey undertaken for a lowland tropical rainforest, Alwyn H. Gentry and Calaway H. Dodson of the Missouri Botanical Garden documented the abundance of various plant growth forms at a site in Ecuador. The number of tree species was remarkable, but the abundance of canopy plants was particularly impressive. The 227 kinds of epiphytes represented more than half of the individual plants and 22 percent of all the plant species in the forest. Adding climbing plants and parasitic plants (mistletoes) to these statistics, about 40 percent of higher plant species in that forest grew in the tree crowns.

Gentry and Dodson surveyed other tropical forests and found an increase in epiphyte richness with increasing annual rainfall. There can be too much of a good thing, however: where rain is often torrential, canopy plants decline. Such downpours may pummel epiphytes, dislodging them from the trees. Other studies indicate that epiphyte abundance peaks at middle elevations in cloud forests such as Monteverde. Jeff Luvall has a tower in one such forest. His data show that conditions in the canopy are more equable, with less searing radiation and more rain and mist than at his lowland forest site. This helps account for the cloud forests' sumptuous vegetation.

Despite their small individual sizes, the significance of epiphytes to tropical ecology cannot be disputed. "Green biomass (and presumably photosynthetic capacity as well) of nonvascular and higher plants anchored in tree crowns can rival—probably even exceed—that of [the trees]," notes epiphyte specialist David Benzing of Oberlin College in Ohio. The plants thereby capture much of the available energy in a rainforest.

The Structural Diversity of Epiphytes

Scanning the Monteverde canopy garden around Nalini and Teri, I noted that the diversity of cloud-forest epiphytes was as impressive as their abundance. I saw myriad botanic forms, ranging from daring sculptural masterworks to what appeared to be slapdash attempts with Play-Doh. Among the species were branch-bound shrubs, squat rosettes, and what looked like membranes plastered against the tree bark. Strewn in the foliage, a pleasing mix of flowers alternated blue and red from one spot to the next. Even the leaves of the epiphytes were blotched with color and were far more varied in size and shape than were those of the trees themselves.

This structural variety is evidence of different strategies for retaining water and the nutrients that water contains. Stress on epiphytes depends on their location among the canopy strata and their position along a branch. In general, the epiphytes with the best-developed traits for capturing and storing

TOP: The *Guzmania musaica* bromeliad—an arboreal plant in the pineapple family—can be found in the western lowlands of Colombia. The gutter-shaped leaves intercept water and impound debris. Several hundred aquatic species, such as mosquito larvae, and uncounted terrestrial ones, including adult frogs, are known to spend at least part of their lives in canopy bromeliad oases; many dwell only within the confines of treetop plants.

BOTTOM LEFT: Fungal bodies large enough to be visible to the naked eye are scarce in the canopy (New Guinea).

BOTTOM RIGHT: Filmy ferns are a mere cell-layer thick (Ecuador).

LEFT: Some forms of the lower plant *Lycopodium* are used medicinally in Sri Lanka, for example to heal bone fractures. The old, spent leafy growths decay below but stay in place, providing sustenance for fresh foliage above.

BELOW: Velvety *Coenogonium* lichens from a New Guinea moss forest canopy are a far cry from the more common rigid, encrusting lichens. Lichens are a combination of fungi and algae; soft forms like *Coenogonium* are mostly dominated by algae.

BOTTOM: Orchids are by far the most diverse group of epiphytes worldwide; many species live on Mount Kinabalu in Borneo.

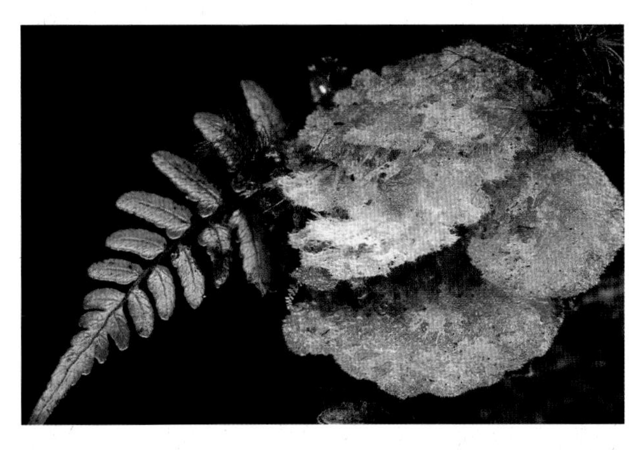

TOP: In a Colombian cloud forest, baby bromeliad plants find room to sprout from seeds in the soil particles on one leaf of their own mother. The little plants are likely to have a short life there: unlike specialized mosses, lichens, and liverworts, bromeliads cannot survive on leaf surfaces.

LEFT: Epiphylls such as liverworts are not altogether a bane for their host. Some contribute nitrogen to their leaf or render it less appetizing for herbivores.

water occur in the highest, most arid forest strata and the outermost, exposed parts of a tree's crown along spaces between trees. It's no surprise, then, that cacti (long and segmented, like green tapeworms) grow as epiphytes in rainforests right to the top of the trees.

Compared with the flimsy leaves unfurled by most tropical trees, many epiphytes invest heavily in each leaf. Near Teri, for instance, I noticed the hard, spiked leaves of bromeliads and the rubbery ones of orchids shaped like squashed fingers. Epiphytes such as these grow leaves with painstaking slowness, perhaps because of the difficulties in obtaining nutrients. Much as occurs in the stems of aerial cacti, the leaves store moisture within succulent tissues, while the waxy cuticles reduce water loss. A leaf may last for years—even for the life of the plant. During that time the tree in which the plant roots may drop and resprout leaves repeatedly. Individual leaves matter less to a tree than to an epiphyte, since nutrients from a tree's fallen leaves can be recaptured by ground roots.

Epiphytes have other tricks to cope with shortages of water and nutrients. Gazing at one of the whorl-shaped bromeliads, I saw a droplet meander down the channel along the central axis of one turgid leaf and slide into the pool at its base. Large specimens hold a gallon or two in the "tank" at the center of their leaf spirals. The tank also hoards debris that falls inside. As the debris decays and releases nutrients, the bromeliad absorbs the broth through special hairs on its leaf base. The tank contents are so bountiful that the roots of some bromeliad species are vestigial and basically useless for nutrient uptake; they glue the plant to the tree.

Some epiphytes form root mats that mop up liquid like a blotter; others have bulbs to store water internally. Orchids and anthuriums develop a velamen, dead tissue around the roots that swells to retain excess water and reduces evaporation from the roots when dry. The velamen sheaths therefore serve some of the moderating roles that soil usually plays.

The ferns and mosses in the Monteverde canopy garden look fragile compared with more robust epiphytes. They grow most abundantly at humid sites or where there is ample mist, as in the half-light of the understory or the shady undersides of branches. Nevertheless, they, too, cope well with dehydration. Filmy ferns, for example, with fronds one cell thick, resemble tattered tissue paper when dry, but revive to full vigor as soon as rain or dew slides from one translucent frond to the next. Ken Clark designed his mist collectors to mimic this action.

The diverse pastel blotches on the tree leaves are also epiphytes of a sort. Called epiphylls, these minute lichens, liverworts, fungi, mosses, and films of algae produce a subtle beauty, with the tonal richness of marble. Epiphylls can also dry out without much mortality.

Most epiphylls would appear to be harmful to the tree. They reduce photosynthesis by blocking light energy from leaf chlorophyll. Indeed, a function of the elongate tip on many tropical tree leaves may be to assure that rainwater drips off quickly, keeping the leaf surfaces dry. Although dryness may not rid leaves of epiphylls, it may impede their spread.

A few epiphylls appear to compensate their host for their use of leaf space: they fix precious nitrogen, some of which they release onto the leaf where the host can then absorb it. Certain bacteria and fungi associated with roots of many plants do the same thing.

In addition to structural features, canopy plants use physiological mechanisms to conserve water. Many epiphytes perform a type of photosynthesis called CAM (Crassulacean acid metabolism), which they share with many fleshy desert plants. In "normal," so-called C_3 plants, tiny leaf pores open during the day to let in carbon dioxide. In CAM plants the pores open at night instead and the plants store carbon dioxide until daylight. Then they, like C_3 species, use light to convert the carbon dioxide and water into glucose. By keeping their pores closed in the midday heat, CAM plants evaporate less precious water than C_3 plants do. But under milder conditions (as in the understory), C_3 plants grow faster than CAM plants because they are more efficient at taking in the carbon dioxide they need for growth.

Development of an Epiphyte Community

After hours of counting epiphytes in one tree, Nalini was still going strong. "There are probably a thousand dollars' worth of orchids right on this branch," she said. Recent economic studies show that treetop products could be more valuable than lumber from the tree itself. Epiphytes include not only ornamentals but medicinal and edible species. Because of the sluggish growth of epiphytes under natural conditions, however, by far the most profitable yields for the plants are obtained in a greenhouse.

Nalini looks at epiphytes not as individuals but as assemblages of plants. "My dream is to say I understand where epiphytes fit into the forest: how they contribute to the flow of nutrients, energy, and materials. To do this I must understand the forest as a whole, but with the epiphytes highlighted. Then I can start explaining things, like why Monteverde has so many epiphytes."

With this grand mission in mind, she cooperates with chemists, microbial biologists, mathematicians, computer programmers, and field experts like Teri on site at Monteverde. "Oh, you taxonomists," Nalini once exclaimed to her husband, Jack Longino, and me. "You can go anywhere with your little field bag, write notes, check the literature, and your study's done. But you just can't do ecosystem work that way. It's a team effort, and it takes equipment. Jack sees all my luggage and shakes his head."

To understand how epiphytes are distributed among the layers of a tree and along branches within layers, Nalini hauls surveying gear into the canopy to map the plants. This new method of collecting data may prove to be the most fruitful yet for tracking epiphyte communities.

The difficulties of gathering data are hard to appreciate from the ground. In large trees, especially in cloud forests, a walk on a major bough is like struggling through a thicket on a precipice. But some perspective is due here: decades or even centuries may have passed to create this extravagant world. There's a history of a million creeping vegetable battles waged by plants seeking to seize and dominate every square inch as disturbances or branch growth opened up new territory. The outcome is not plants heaped randomly one atop another but a landscape of an elegance unachievable with the modest geometries of the human gardener. This is the supreme physical manifestation of the rainforest tapestry.

Tree branches elongate by growth at their tips. What this means, Nalini explained, is that the epiphytes near the fork of the limb have had the most time to accumulate; those near the twigs found lodgings only recently.

Proceeding along the branch from twigs to trunk is like traveling through time. At first the bark is clean of soil. The smattering of epiphytes survives on dust and dew; they are often relatively exposed and so must contend with extremes of temperature and moisture. With passing time, soil thickens. The underside of the branch still suits only a few species (primarily mosses and lichens), but the sides and top, where soil and light are plentiful, harbor plants in ever-increasing diversity.

In the most ancient neighborhoods of the canopy, near the base of large branches, the soil can reach a half foot or more in depth, and may be deeper and richer than that of the forest floor. The number of plants for a given surface area can be far greater than on the ground. Bulky epiphytes appear, even woody forms such as shrubs and small trees.

The sequence typically stops just before the branch splits off the trunk. Tree crotches generally have deep, moist soil but seldom harbor many plants; most may be too heavily shaded. The roots of epiphytes adjacent to the crotch tap into this soil, which may remain damp even during droughts. (When the ground is dry, many animals and human explorers climb trees to search for drinking water in the forks of the branches.)

Fungi, however, prosper at tree crotches, although conspicuous fruiting bodies, such as mushrooms or bracket fungi, are nowhere as common in the canopy as on the ground. Most fungi occur as a spider web of strands (hyphae) in soil mats and rotting wood.

A corollary of this chronology is that epiphytes contend with ever-shifting conditions during their lives. An orchid on a sunlit twig will be overshadowed as the surrounding foliage grows. As its twig expands into a bough, competition with other plants for space intensifies.

FACING PAGE: The flaky bark of the "naked Indian tree" (*Bursera simaruba*) prevents epiphytes from accumulating.

Within broad limits, most epiphytes don't care what tree species they live on. Nevertheless, physical traits—girth and inclination of branches, texture of bark, stratum in the canopy—influence the prospects of different epiphytes on different trees by aiding or impeding the attachment of the plants and the accumulation of canopy soils for them to grow on. As yet unexplained variations in the chemical composition of canopy soils from tree to tree can also affect the epiphyte communities.

Certain trees support virtually no epiphytes. These include palms and other fast growers, those housing ant colonies, as well as those with bark that sheds, sloughing off epiphytes. The so-called naked Indian tree of tropical America and Florida (*Bursera simaruba*) is one of the latter, named for its thin, flaky, skin-colored bark.

Possibly such trees denude themselves for a reason, leading us to question the desirability of epiphytes from a tree's point of view. Epiphytes weigh down trees and so may raise the likelihood of breakage. Epiphyte shape or volume may reduce tree streamlining, and so increase the damage inflicted by wind.

David Benzing thinks of epiphytes as "nutritional pirates" that remove nutrients before the tree can take them. This may be accurate in lowland rainforests, where tree-crown roots are rare. Nalini Nadkarni's view of cloud forests is that epiphytes intercept nutrients that otherwise would never have made it into the ecosystem at all. Because trees with canopy roots take advantage of these nutrients, epiphytes may actually invigorate their trees.

Nalini emphasizes the positive contributions of epiphytes to the trees. She notes that branches grow in concert with their epiphytes, and so should develop the tensile strength to support them. Still, broken branches and snapped trunks appear more commonly in the New World than the Old (not a trivial fact, since the size of forest gaps has repercussions for tree communities: see Chapter 2). Bromeliads, found only in the New World, could be an added burden to the trees there. In one Colombian cloud forest, water held by the bromeliad tanks alone amounted to more than twenty tons per acre!

Brainstorming the Science of Epiphytes

The Marie Selby Botanical Gardens in Sarasota, Florida, holds the largest scientific and public collections of epiphytes on earth. It hosted a symposium on the biology and conservation of epiphytes in May 1991. Nalini Nadkarni, at that time the Selby Gardens' director of research, organized the event with her usual boundless zeal.

Scientists attend symposia for many reasons: to learn techniques; to argue with colleagues; to develop collaborations; to find out what others in their discipline are doing. Particularly for those just beginning their scientific studies, symposia are opportunities to meet people with similar interests outside their university. To students, the most famous experts in their field take on an almost mythical stature; they are the shamans of that science's tribe. When I arrived at the conference, I could identify these individuals by the tight gatherings around them.

Horticulturists have fancied epiphytes for centuries, but the study of epiphyte ecology has barely begun. In Sarasota, biologists from many countries contributed provocative research results on everything from newly described species to tribal uses of epiphytes in the Amazon. Ariel Lugo and Frederick N. Scatina of the U.S. Forest Service argued for using the health of delicate epiphytic species to assess global climate change. Darwyn S. Coxson of McMaster University in Ontario described how rain leaches sugars out of the tissues of canopy moss, yielding sustenance for other plants—

another pathway by which nutrients cycle through the canopy.

With Jeff Luvall's tower on my mind, one presentation in particular caught my attention. Stanley Herwitz, a forest hydrologist at Clark University in Massachusetts, described his perplexity with the dynamics of tropical rainfall. In northeast Queensland, Australia, he had combined his estimates of the amount of rain striking the forest floor as free-falling drops ("throughfall") with those of rain draining down tree trunks ("stemflow"), and measured huge differences in total ground-level precipitation between one tree and the next.

To see if this was due to differences in the canopy's absorption of water, Herwitz hung freshly cut trees in a room where water dripped from a bank of syringes. He found that tree species differed sharply in the amount of water retained by their crowns, but not enough to account for discrepancies in his field data.

Then he noticed the forest's "roughness": variations in tree height and shape that make the canopy uneven. Indeed, by taking precise readings of canopy height from Panama's canopy crane, Alan P. Smith of the Smithsonian Tropical Research Institute in Panama and his colleagues have shown that the roughness of species-rich tropical forests is far greater than that of temperate-zone forests and is mostly concealed from ground view. (I find this easiest to appreciate from an airplane, especially over the dipterocarp forests of Borneo.)

Since winds blow consistently from one direction at Herwitz's study site, tall trees cast predictable rain "shadows" on smaller trees to their leeward side, where moisture could be critically low. Through computer modeling he learned that when one inch of rain descends above the canopy, some tree crowns receive half an inch of water, some an inch, others three inches or more. It is possible that an epiphyte could be on the verge of drowning on one branch while another suffers dehydration only yards away. Conditions can be far less predictable in a tropical forest than they might appear.

Staghorn ferns can reach the size of a bushel basket. Their centermost leaves die and decay, forming compost for the plant itself (Papua New Guinea).

Conversations led to further questions. Why is Latin America so rich in epiphyte species? (Possibly because the Andes Mountains have provided the large expanse of cloud forest in which new species can evolve.) Isn't it odd that in much of tropical Asia ground space is dominated by one group of trees (dipterocarps), but no epiphyte group dominates the canopy? Yet in tropical America, canopy space is dominated by a single group of epiphytes (bromeliads), but no tree group dominates the ground. (No one could explain this.)

The buzz of ideas felt reassuring: the study of epiphytes was booming. Perhaps someday this special branch of canopy biology will help us manage and conserve the delicate tropical environment.

CHAPTER FIVE

TAPPING THE GROUND

Upon first entering the rainforest in Borneo, I was surprised that palms I'd seen overrunning the trees had long, inch-wide stems instead of stout trunks. These rose in messy spirals, held to tree trunks by strap-like extensions that tore my arms and legs with spines as nasty as sharks' teeth. Walking off the beaten path, I blundered into a maze of coils that resembled the barricade of barbed wire around a prison. Bloodied, I sank knee-deep into mud. Above, I could see how the palm's stems looped out of view into the canopy, spines ratcheting against tree limbs along the way. Eventually, I pulled one aside and made my escape.

Picking at my rice beneath a dripping lean-to later that day, I learned from forest guards that these climbing plants were rattan palms, valued for furniture. Rattans use their spines primarily to claw their way into the canopy and secondarily to keep entomologists and other large vertebrates at bay. Perhaps my scars from this encounter diminished any aesthetic appreciation I could have been nurturing for climbing plants. In northern latitudes, climbers, like ivy or grape vines, are easily overlooked; in the tropics they form a deluge. As Polish botanist M. Raciborski wrote of arriving at a tropical rainforest in his book *Life on the Equator*:

> We stand wonderstruck and helpless; without a man-made path there's no access into the forest. Its margin is like a wall bound by lianas in all directions and covered with leaves of every possible hue of green. Natives are sent ahead to cut a path with handy heavy-bodied knives called machetes, while we rest in the shade. The lianas are most impressive . . . entwining the trees in countless, irregular coils in search of light in the canopy of leaves.

I admit that my own experiences led me to look at vines and lianas as nuisances rather than as sources of inspiration.

Lianas and Vines

Years afterward, while downing beef stew in the more refined atmosphere of the cafeteria on Panama's Barro Colorado Island, I voiced my opinion of climbers in front of several resident biologists. This raised the hackles of botanist Darlyne Murawski, who proposed to set me straight. I promptly found myself rambling the island with Darlyne, who introduced me to climbers at every turn.

This small (six-square-mile) island harbors 175 species of climber. Given that all climbing plants must be much longer than they are wide, an aspect of climbers that I hadn't been prepared for was their

PAGE 82: Rampant hemiepiphyte roots and lianas girdle a tree trunk in Corcovado National Park, Costa Rica.

PAGE 83: A vine in Panama reinforces its grip with tendrils that cling to each other if all else fails.

FACING PAGE, TOP: In twiners, the whole plant wraps around a support, and in time may distort the growth pattern of the host, sometimes even killing it.

FACING PAGE, BOTTOM: In *Macfadyena* highly modified compound leaves extend at intervals: part of each leaf forms a three-pronged hook to grip the supporting trunk. Though climbers evolved from trees and shrubs in many plant groups, most have opposite leaves, which branch off directly across from each other. This gives them a better hold than alternate-leaved plants, for which climbing would be as challenging as it is for a one-armed human. Some intermediate strategies exist, such as tree-like plants too weak to stand on their own.

TOP RIGHT: Scramblers grapple onto supports loosely with spikes. Some seem to knock trees down as part of their normal life histories, thereby creating their own gaps. BOTTOM LEFT: Many aroid plants such as *Philodendron* encircle trees with modified lateral roots.

TOP LEFT: In tendril-bearing species like *Omphalea*, lateral tendrils contact a support (in this case a sapling tree). ABOVE: Within two days a tendril has fastened tight. Both twiners and tendril-bearing plants frequently attain the canopy. Some species have leaves that overtop and thus shade the crowns of upper canopy trees, whereas others have leaves that intersperse with or underlie those of the tree.

visual diversity. Though their species are fewer, climbers belong to twice as many families of flowering plants as epiphytes do.

Darlyne pointed out species new to me, such as climbing ferns, climbing bamboos, and climbing gymnosperms (conifer relatives). For their assault on the canopy mountain the plants came equipped with whips, claws, hooks, glues, Velcro-like bristles, pitchforks, suckers, corkscrews—seemingly in every combination. Structure varied as well. Many of the bigger climbers were plaited like rope. A tree's wood is built for rigidity; a climber's for flexibility and long-distance water flow.

Among the island's climbers, my favorites were flat ones that looked like ribbons. In South America I had come across such "monkey-ladders" a foot broad by a half-inch thick, with elegant undulating surfaces. Where a loop of one monkey-ladder hung near the ground, Darlyne and I could not resist sitting on it in turn and, kicking with our feet, swinging like schoolchildren.

Both delicate and robust climbers abounded. One specimen, a member of the pea family with a cylindrical bole two feet in diameter, was Jack-in-the-Beanstalk come true. Initially shooting to the heavens like a tree trunk, it subdivided again and again to create a viny cosmos, swathing sixty-four canopy trees spread across an acre with leaflets so fine that it was hard to believe they belonged to such a large organism. This plant, the research subject of Jack Putz at the University of Florida, may be the most enormous climber ever studied. The total length of its bole has not been calculated, but other climbers around the world have measured in at over half a mile and weigh in the hundreds of tons, more than most of the trees supporting them.

These titans also sprout more foliage than any single tree. Jack calculates that a climber three inches wide may have as many leaves as a tree with a trunk two feet wide. The difference is that the climber, freed from supporting its own weight, can direct more of its resources into dominating the canopy. Trees of a given height must have trunks that exceed a certain diameter to support their weight; climbers have no similar size limitation.

With their ground connections, climbers can withstand conditions that would be too dry for many epiphytes. But transporting water and nutrients over long distances requires special anatomy. The vascular channels of most plants are undetectable, but in climbers they are wide enough to be seen as pores when the plant is cut. Large quantities of liquid pass through with ease. In temperate climates such wide vessels would be detrimental; freezing can introduce air into them, and air, blocking nutrient and water flow, can be as deadly as nitrogen bubbles in the capillaries of a careless diver. This may explain the scarcity, except for some ivy species, of long-lived woody climbers (called lianas) in the temperate zones. Short-lived herbaceous climbers (called vines) are common in the understories of both the temperate and tropical regions. The paucity of climbers in tropical cloud forests, however, remains a riddle. Freezes don't normally occur in cloud forests, and other canopy plants flourish in them.

A thirsty explorer can hack a yard's length from certain lianas and attempt to drink the blood-red, milky, or clear fluid that drains from its vessels. "But don't try this needlessly," Darlyne admonished. Seldom can so magnificent a creation be slain with such ease: the limited segments visible from earth may suggest that lianas are insignificant, but in fact many plants are incredibly far-reaching and very old.

Almost any part of a plant can be modified into a contrivance for scaling vegetation. "Darwin wrote a book on this subject," Darlyne reminded me, clinching the issue of the importance of climbing plants. Some species use tendrils—projections evolved from leaves or branches. Tendrils sprout at intervals along the stem and curl around anything they touch. They can be sensitive. At the turn of the century, a biologist showed that whereas tendrils ignore ordinary dripping water, they twist dramatically in response to water droplets that contain a fine powder! Moreover, tendrils of some plants explore preferentially in the direction of nearby objects, apparently sensing them (as if by smell) from afar.

Other climbers, called twiners, wind their entire length around a support plant; the tip revolves as it grows. Still others use modified roots or twigs to adhere to the trunk or grip cracks in bark. Finally, scramblers simply grapple their way upward through foliage without clinging tightly to any part of it.

Each strategy is ideal for certain circumstances. Tendrils work best with slim supports. Twiners, perfect for supports of moderate diameter, fall into a heap if they attempt too wide a limb. At the other extreme, adhering roots succeed with trunks too bulky to be held any other way but fail with slim vegetation. This strategy also makes it impossible for the plant to switch from tree to tree.

Scramblers need dense vegetation to hold them up. To illustrate this for me during a canopy climb, Darlyne enthusiastically pulled at a thorny scrambler that meandered toward us from an adjacent tree crown. When that tree swayed away from us in the breeze, she let out a shriek. The cords of foliage snapped out of her grip, and the scrambler slid back ungracefully among the branches. Darlyne's fingers had been bloodied by its recurved thorns. It's no accident that this type of plant remains little studied.

Lianas start life indistinguishable from seedling trees. These young grew everywhere Darlyne and I looked. Wherever *Doliocarpus* juveniles were exposed to sunbeams in small canopy gaps they transformed, thrusting upward and acquiring a long

The specter of a tree that once stood at a site is hinted at by the strange form of a liana in Amazonian Ecuador.

terminal shoot. "Unless a young vine finds something to lasso, it eventually collapses under its own weight," Darlyne remarked. Fallen individuals continue to send up shoots one after another, in search of that crucial first toehold toward the canopy.

"Tendril-bearers and scramblers need frequent stepping-stones to continue skyward," Darlyne said. "Stepping-stones" are twigs and branches of other plants. Most climbers cannot bridge a space of more than one to five feet between these support rungs. The first steps upward can be difficult in the relatively open understory of an undisturbed forest. By contrast, any place with a thick understory (riverine forest, hurricane-prone forest, or a major treefall gap) provides support every step of the way. Such "jungles" acquire a riot of vines at ground level, if not impenetrable bordering curtains of climbing vegetation. As naturalist-explorer Peter Matthiessen recounts, "In New England one walks quite gradually into a wood, but not so in the jungle. One steps through the wall of the tropic forest, as Alice stepped through the looking glass; a few steps, and the wall closes behind."

Pristine tropical forests may be as laden with climbers as any of these heavily disturbed places. Their greenery is just not visible from earth, having grown over the top of the canopy. In New World rainforest (where most research is based) about half the canopy trees have lianas. These either reached the canopy long ago, during periods of forest disturbance, or ascended tree boles or other lianas.

When a liana arrives at a tree's summit, it extends toward the heavens before falling back to the canopy surface. Continuing in this clumsy fashion, it grows horizontally across the crown. In many cases the plant expands over swaths of canopy by repeated forking. Eventually it comes to the crown edge, where it must contend with crown shyness if it is to spread further.

To bridge a gap between trees, a terminal shoot arches up, then leans over to the next crown. This is accomplished through trial and error, so that shoots may sprout at different angles before one succeeds. The initial connection is hard to maintain. Crowns often throw off tendrils while swaying in the wind. Gaps wider than a couple of yards cannot be bridged at all.

In attempts to cross an abyss all or part of a liana may plummet over the crown edge. The plant can still succeed in an assault on the adjacent tree. Now it does it the hard way: starting from below, it locates and mounts a suitably proportioned trunk.

Growing slowly in girth, lianas may outlive many of the trees that support them. Indeed, they are built to survive smashing and twisting, and they seem potentially immortal. Yet forest upheavals leave permanent marks on lianas. Climbers spiraling in empty air, odd shifts in direction, or distortions in form—all suggest the ghosts of trees past. As nineteenth-century explorer Alfred Russel Wallace wrote in a book of his impressions of the tropics:

> They twist around the slenderer stems, they drop down pendent from the branches, they stretch tightly from tree to tree, they hang looped in huge festoons from bough to bough, they twist in great serpentine coils, or lie in entangled masses on the ground.

For example, a liana draped partway to earth between crowns tells us that a tree bearing it aloft once stood at the intermediate spot. If we better understood the language of these living cables, we might read the local history of the rainforest in their convolutions.

Climbers drive some of this history themselves. Lianas are a burden to trees, spreading over and shading them. They slow growth and even snap saplings they use as stepping-stones. Trees improve their chances of surviving if they can grow faster than the climbers on them. Jack Putz has found that trees with large leaves, shedding bark, and swaying trunks often dislodge pesky climbers. Fewer climbers may be able to bridge trees that sway out of sync, he notes. Perhaps the varied architectures of tropical trees result in different rocking patterns. Jack conjectures: might this promote local tree diversity?

Drooping languidly over trees, lianas appear to have little of the vigor needed to reach the canopy. Sometimes even botanists assume that lianas are "carried up" like hitchhikers on a maturing tree. "Absolutely not," Darlyne insisted. Because plants grow only at their tips, any points along a trunk forever remain the same distance apart; growth there occurs only in girth. A liana then, has to do all the climbing by itself. And an impressive job it does: some can lengthen a half-foot a day.

The tenacity and strength of lianas influence forest dynamics. They entangle trees, so that a tree thrown over by wind is likely to pull down its neighbors. With their springy, well-padded tissues, lianas almost invariably survive and perhaps even encourage the fall to continue their interminable spread.

In the Congo's *Gilbertiodendron* forests, an extravagance of sprawling lianas have attained sizes as imposing as pythons. In adjacent forests with a high tree diversity, lianas of similar girth are scarce. Perhaps lianas have a stabilizing effect on monodominant forests because wind acts with little effect on their uniform, and thus relatively streamlined, canopy surface. Braced by lianas, even weak trees can remain upright. The rarity of cataclysms in these forests may permit the lianas to reach a prodigious size and age.

Parasitic Plants

Cristián Samper, then (1990) a graduate student at Harvard—he now works for environmental agencies in Colombia—walked briskly ahead of me through the botanical paradise of Colombia's La Planada Reserve. As we splashed across streams and wallowed over hills oozing mud, he showed me canopy wonders. Despite pleasant temperatures at this 5,500-foot altitude, my sweat had turned me into a musty sop. Cristián looked as cool, neat, and unflappable as a young businessman on a stroll downtown. He pointed again to the canopy. "That parasitic plant—*Oryctanthus*—it's an odd one," he said. "At first it forms a shrub on a branch of a tree, like any mistletoe. Then it extends a shoot from branch to branch, like a vine, plugging into the tree's vessels each time. It even leapfrogs between trees."

Climbers, hemiepiphytes, and parasites exploit

The sprouting tips of mistletoe seeds plug directly into the tree's bark (Colombia).

trees for support. But unlike epiphytes, they maintain some sort of connection to the ground, which remains their primary conduit for water and nutrients. Of these plants the parasites are most devious. Parasites drink from the earth indirectly by tapping the vascular system of their host tree. They take sap much the same way a tick, embedded in skin, drinks human blood.

There are about two thousand species of parasitic plants. Terrestrial ones, feeding from the roots of trees or other plants, have mostly forgone green photosynthetic pigments entirely. Their characteristically pale or red stunted forms appear luminous against dusky leaf litter—if any part of their body lies aboveground at all. Everything they need comes from the host, including sugar transported from where it was photosynthesized in the host's leaves. The parasite provides nothing in return. The most famous species, *Rafflesia arnoldii*, the stinking corpse lily of Sumatra and Borneo, develops monstrous rust-red flowers a yard wide. The handsome blooms send humans gagging from their stench, but to the flies that mistake them for rotting meat and pollinate them, they are ambrosia.

Canopy parasites, Cristián told me, are universally referred to as mistletoes, though they belong to several independently evolved groups, all of them woody. People associate mistletoes with northern latitudes because of holiday traditions. In fact, most are tropical. Rainforest mistletoes often sport pretty flowers, in contrast to their somber temperate cousins.

Mistletoes often appear as dense patches of shrubbery within a tree's crown (Colombia). *(Photo: Darlyne Murawski)*

Most mistletoes have greater fidelity to their hosts than do epiphytes, but usually occur on several tree species. Not completely parasitic, they retain green tissue: like ordinary plants, they photosynthesize their own sugars. Indeed, mistletoe leaves may actually resemble those of their hosts, making some species hard to pick out from a distance except by subtle changes in color and texture.

Birds defecate mistletoe seeds on branches (see Chapter 9), where they often germinate within hours. The young plants live as epiphytes for up to a year until they penetrate the branch, forming a messy fused ball-in-socket joint within the wood. It is in this buried swelling, called the haustorium, that a mistletoe siphons nutrients and water from the host's upward flow of sap. In effect, mistletoes use the tree as a substitute for roots of their own.

Parasites do obvious harm to trees because they draw directly from the identical pool of resources as the branch on which they grow. In fact, occasionally the part of the branch extending beyond a mistletoe's point of attachment dies of thirst.

Hemiepiphytes: Starting from the Top

Cristián Samper researches hemiepiphytes —plants that are epiphytes for part of their lives—at La Planada Reserve. A mile from where we walked beneath mistletoes and epiphytes a splendidly organized field station virtually unknown to foreign scientists sits against a sweeping Andean backdrop.

La Planada lies in the greatest expanse of cloud forest on earth, a vast breeding ground for new plant species. In disturbed forest around our quarters, epiphytes that anywhere else would be hidden in treetops adorned boughs at eye level with garlands of flowers. In deep forest, plants formed herbaceous

FACING PAGE, TOP: Cristián Samper, an ecologist studying at La Planada Reserve in Colombia, examines the initial descending root of a *Clusia* hemiepiphyte—its large, leathery leaves red on the undersides—growing in a patch of canopy plants next to him. A root may lengthen by a foot or more a day, a rate of growth matched by a few vines and pioneer trees. Once a root strikes ground the prospects for the plant's survival are good. BOTTOM: Cristián looks up at a lopped-off branch suspended by a pulley from an epiphyte-laden tree bough. Standing on the ground, he can raise and lower the branch segment to study the young *Clusia* plants growing on it. Two roots of an adult *Clusia* plant descend to the left. Incidentally, swinging on vines as if they were ropes seldom works: a liana extends continuously from its rooted, earthbound end to its farthermost growing tips, and these dangling tips are usually too slender to carry weight. Tarzan would probably swing from sturdy roots of adult hemiepiphytes lopped off at the base.

coats a yard thick on trunks and hung like draperies from branches. The vegetation of any other cloud forest I had seen paled by comparison.

We stopped to look at the roots of the type of hemiepiphyte studied by Cristián, *Clusia*. A half-inch or so wide, the roots dangled straight from the canopy to the ground at our feet. They felt as taut as strings on a supersized banjo. I plucked one and watched it oscillate in waves up to the canopy, where I spied, at last, the big rubbery leaves of the same *Clusia* plant.

By definition, a hemiepiphyte switches survival strategy over its lifetime. Living first as a canopy epiphyte, a juvenile *Clusia* may grow just two leaves a year. In a few years it will have accumulated sufficient reserves to grow a root abruptly to earth. No longer impeded by the canopy's unpredictable water supply, the plant can now reach a huge size by sending down more roots. Eventually it may dominate the crown space of the host.

"Hemiepiphytes are like lianas," Cristián explained. "They do not need much structural support, they just hang there." In his *Clusia*, for example, the similarities run deep: as in lianas, *Clusia* roots have extra-wide plumbing to transport water swiftly.

Clusia seeds sprout so quickly that you can almost watch the seedling unfold, Cristián told me. "They are not choosy about where they germinate. But for some reason a seedling requires a branch surface to root properly. A dead branch on the ground will do. Maybe the plant will grow a little faster in bright canopy light, but the point is, unless it's on a branch, it will *die*."

Why don't some end up as ground plants? "Mechanical damage," Cristián said. "A seedling can be killed by just one leaf hitting it from above." Cristián has hung branches harboring young *Clusia* from trees, using a pulley system. Lowering the branches periodically, he checks the fate of the plants.

To survive long, *Clusia* must germinate not only in the safety of the canopy but about halfway out along an intermediate-sized branch. "That's different from figs, some of which are also hemiepiphytes," he continued. "Figs survive most often at the crotch of a branch. Everything has a pattern. I have noticed that mistletoes here usually attach to the tree beneath a branch." Apparently all treetop plants, not just epiphytes, must find a distinctive foothold in the densely packed canopy community.

Why the habitat differences? "Maybe seeds get deposited by animals and then are washed elsewhere—I think this may happen with mistletoe seeds—or perhaps it's due to shifts in humidity or predation from place to place. We can't say," he admitted. "To study germination, you have to think on a fine scale. Most people worry about big things, like where a whole tree is."

The most fantastic hemiepiphyte strategy, used by some *Clusia* and many figs, is that of the stranglers. Starting as an epiphyte, a strangler extends its roots down the trunk of the support tree, plastered against the bark. Whereas the roots of other plants remain simple conduits to the earth, in the extreme

Mature *Clusia* fruit open (dehisce) like flowers to reveal orange-red flesh that is devoured by birds (Colombia).

FACING PAGE: An adult strangler fig in Monteverde, Costa Rica, makes a superb climbing toy—a definitive means of reaching the canopy unencumbered by human gadgetry. At the tree's base, the climber can slip between the roots that form the tree's trunk-like shell, then grasp a root from inside as if it were the rung of a ladder. A long-dead host tree once grew in the center space. Above, light beams shoot across the vertical tunnel through hundreds of portals. The portals reveal the outside world, bit by bit, step by step, as one ascends.

forms of stranglers the roots coalesce until they form a cramped basket around the tree trunk. Then, as the tree attempts to grow, they crush it to death.

This description is too colorful for some biologists, who downplay such notions of tree strangulation. They insist that trees die by the overshadowing of dense hemiepiphyte foliage, or by competing with the roots of the strangler for nutrients and water. I find it odd that this vegetable version of nature-red-with-tooth-and-claw has been watered down. Undoubtedly, the roots of a strangler fig impede expansion of the tree. Inside their vise-like grip, the trunk bulges from each opening in its woody cage. Such expansion is absolutely necessary for the tree's survival. In the trunk, new vessels must periodically replace dying ones. If rings of vessels cannot be added by increasing trunk girth, nutrients and water cease to flow as surely as a tourniquet stops blood flow to a human limb. Jack Putz agrees with me. "That's why strangler figs seldom kill palm trees," he explained. "Palms don't need to grow in diameter to replenish their vessels."

The peepul tree (the species under which Gautama Buddha attained enlightenment) seems especially malicious. Starting as an epiphyte, it drives roots into the host, literally tearing it apart.

There's no dispute about what happens after a strangler kills its host. The tree corpse, cradled aloft in the arms of its slayer, rots and falls away, dusting passersby below and making compost for its replacement. And a replacement it is: what was once an innocuous canopy plant now stands on its own right as a tree, albeit a curious one. Its trunk persists as a woody cylinder, hollow down the middle where its predecessor once stood. Continuing to grow to an enormous height, the trunk retains countless nooks and crannies, left over from the fusion of scores of roots. These form wondrous hideaways for epiphytes and canopy creatures.

What a wonderfully wicked strategy the strangler uses! It circumvents the indignity of being one of the struggling saplings in the gloomy understory to become a major class of canopy tree.

Secondary Hemiepiphytes: Starting from the Bottom

Another class of hemiepiphytes turns the life history of the species discussed thus far on its head. Unlike primary hemiepiphytes like *Clusia* that begin life as epiphytes perched in trees, these start at ground level as vines and often work their way up to becoming epiphytes. Since they transform into epiphytes secondarily, we call them secondary hemiepiphytes. In their own way as flamboyant as any strangler, secondary hemiepiphytes wander through a forest, looking and acting much like a snake in search of a place to bask.

Surprisingly, these plants belong to groups that we regard as mundane: houseplants such as philodendrons and monsteras imported from the American tropics. Even specimens in dentists' offices need a reminder of their arboreal origins: they climb by wrapping modified roots around the artificial tree trunks in their pots.

University of Delaware professor Tom Ray, as a graduate student at Harvard, studied monstera seedlings scattered in terrestrial soil. He found that they initially do an odd thing: ignoring the light they need for survival, they head toward the nearest dark object. This strategy usually leads them to a tree trunk. Once there, they shift to the conventional plant strategy of growth toward light, and so they climb. Eventually the vine's terrestrial roots become superfluous and its stem dies at the tree base. With all connections to the earth

FACING PAGE: A relative of the *Cecropia* pioneer trees, the Ecuadorian *Coussapoa* is a hemiepiphyte and a strangler. But, since palms do not increase in girth as they grow, in this case the *Coussapoa* is unlikely to kill its victim by strangulation.

Monstera seedlings sprout on the ground and orient toward the base of a tree to start their ascent into the canopy (RIGHT, TOP). In the shade of the understory, the leaves lie flat and overlap in a herringbone pattern, but where the plant is drenched in sunlight, the leaves grow ever more expansive (LEFT). The plant casts off its original roots, but may send down new, secondary roots in their stead molded to the surface of a tree's bark (RIGHT, BOTTOM) (Panama).

severed, the plant potentially can survive henceforth as an epiphyte.

Monstera's foliage also changes radically. In the understory the heart-shaped leaves are a couple of inches wide and often lie flat against the bark. But where the sun shines on the plant, the leaves grow "monstrous"—as much as six feet in length, with deep lobes.

Retaining a stem a few yards long, the most mobile secondary hemiepiphytes mosey through a tree's crown: when in ideal, sunny spots, they move leisurely and unfurl big leaves; but when in shadow, they move quickly by building more stem than leaves. Progression results from the stem growing in front and dying behind: this is the closest any terrestrial plant comes to taking a stroll.

Such plants enjoy unusually flexible lives. Most epiphytes die if they fall from a tree. If a whole monstera (roots and all) takes the plunge, it simply makes for a nearby trunk and heads up again (or finds itself in a dentist's office). If the plant cannot find a route to the topmost trees, it may branch. One section continues to search within the understory. The other returns to earth to seek out alternative trees.

These green snakes have often baffled me. On the one hand, epiphytes live at the brink of death from dehydration or lack of nutrients, the price they pay, it appears, for not being rooted to bountiful Mother Earth. On the other, we find hemiepiphytes that supposedly cast off their earthly connections at the earliest convenience as so much excess baggage.

But it's becoming increasingly clear that secondary hemiepiphytes are more flexible than they have usually been made out to be: at times of water stress, they have the option (unlike epiphytes) of dropping new roots to the ground. Indeed, because such roots, being small and plastered to tree bark, are next to impossible to trace to earth, it may be that many secondary hemiepiphytes have them. Such plants would behave more like vagabond lianas than epiphytes.

Experiencing Time Like a Plant

The rainforest tapestry changes constantly. Our problem as humans contemplating the canopy is that we are oblivious of the movement of plants. What motion we witness belongs to the animals. One thing I've learned from climbing-plant aficionados is that there is art in plant motion. We may appreciate the grace of animals as individuals (the whir of a hummingbird, leap of a monkey, scramble of a beetle), but viewed in reference to the forest, animals move as simple projectiles: dots in a greater framework of vegetation.

Plant dynamics are another matter. Rooted in place, plants give us a dance of interwoven forms that begins just at the limit of our perception of time and proceeds across human generations. There are a few curiosities—sensitive plants or carnivorous forms capable of swift but simple action—but these contribute little to the canopy superstructure. Straining hard, we barely make out the rotation of a climber's terminal shoot, its speed at best matching the crawl of a clock's minute hand.

Accelerate time a hundredfold: about two hours pass in an interval we experience as one minute. Most animals progress in jerky blurs, scarcely identifiable; even lethargic snails sprint. Plants now inspire our awe. The climbing shoot that we had watched before now spirals, like the yard-long feeler of some monster in a Japanese horror film. All around us these shoots rise pugnaciously, groping like blind predators would for their quarry. On other plants, tendrils encircle trellises in graceful arcs, like monkeys coiling prehensile tails; *Clusia* roots glide slowly downward.

Speed time once more by a hundredfold: about one week passes during the single minute of our observations. Animals shimmer in the air, barely registering to our senses; one night's sleep takes a couple of our seconds. Plants have gone wild. Leaves pop open; most flowers unfold and die as a mere flash of color. Climbers move in a combative frenzy, tendrils and twining shoots spinning in a blur; plant tips seem to flow doggedly upward, often piling one atop another in a carnal race to the light.

Now, again, a hundredfold acceleration: about two years pass during our minute. Sapling vines and trees appear as crowds of writhing vermin that continuously die and decay but for a lucky few. Branches of the mature trees overhead thrust outward. Lagging close behind, epiphytes burst forth. Monsteras slither among them uncannily like green serpents, transforming light into motion with absolute silence. In a minute we may be able to detect the ever-tightening grip of a strangler around its victim.

In our next time-jump, the strangler grows its clasping tentacles like . . . well, like nothing that exists in our time frame, but our minds conjure a horrific meshwork of contracting muscle. Trees ramify tumultuously, seizing canopy space in an exuberant dance. All around us during our minute of observation (that is, in two centuries of actual time), collapsing trees tear holes in the canopy that pioneer trees and vines swiftly invade; we see forest ecology in action.

Any further jumps in our perception of time and everything but the geological vista will be blurred. Unable to distinguish even the most enduring organisms, we will have left behind the realm of individuals and entered the domain of biological evolution.

CHAPTER SIX

INSECTS ON A RAMPAGE

One morning while I was climbing in Peru, my support rope shifted and I abruptly fell several inches and began to spin in place. Soil and plant bits shaken down from the branch above whirled into my eyes, blinding me. My hands were full of gear, so to stabilize myself I wrapped my legs around a branch high to one side—and unwittingly smashed an epiphyte garden that concealed a well-defended ant nest. Ant workers descended my leg and fell like kamikazes on the rest of my body. "Camponotus femoratus," *I managed to think as they gashed at my skin with their mandibles and sprayed formic acid into my wounds. I had found my first ant garden the hard way.*

Thirty million. This number has earned my colleague Terry Erwin quite a reputation. Based on his research on tropical rainforest canopies, he estimates that this many arthropod species exist globally—quite a jump from the 1.4 million species, from protozoa to trees, described by taxonomists since Carolus Linnaeus developed our current system of classification some 230 years ago. Indeed, one may argue that the supreme accomplishment of biological evolution is not man, nor his vertebrate kin, but arthropods—and especially one group of them, the insects. As ecologist Robert May puts it: "To a rough approximation, and setting aside vertebrate chauvinism, it can be said that essentially all organisms are insects."

While insects as a whole command much of the earth's surface, many individual species seem to be rare. Rarity may take several forms. A species may be limited to a small area, or be found in an uncommon habitat, or its individuals may be widely scattered. Unfortunately, for most tropical insects, entomologists can't say which, if any, of these descriptions apply: they only perceive that the species are hard to locate.

Occasionally a supposedly rare insect turns out to be common—we simply had not known where (or when) to look for it. In my own studies on ants, I've occasionally solved puzzles of this sort, as when I discovered "rare" *Recurvidris* ants during what I intended to be a brief tour of a Balinese temple. Only one species in this genus had been described until then, but to my surprise two new species thrived in the forest near the temple. For days an incessant stream of tourists enjoyed the novelty of watching me set up experiments with sugar, forceps, magnifying glass, and other odds and ends. Once I understood where to find these ants, I tracked down more with ease.

In a rainforest, searching for a particular little-known organism like an ant can be impractical—a question of luck—regardless of how common it may be. Rainforests seem to offer infinite hiding places. A precious pinned specimen may be accompanied by just the vaguest information, often in the antiquated scrawl of a long-deceased explorer: habitat data (perhaps "on trunk"), along with an inkling of its locality (perhaps "near Bangalore").

Probably because of this, rarity isn't the first thing to leap to the mind of most entomologists musing over tropical insects, as it might be for botanists pondering massive tropical trees. It is the diversity of novel forms, colors, and sizes of insects that staggers us first and foremost.

By suggesting that there may be so many species, Terry Erwin's major accomplishment has been to draw attention to a neglected habitat for biological diversity: the rainforest canopy. Trees may be the ultimate tropical hiding place. Until Terry's estimate, few scientists bothered with tropical treetops. Suddenly the canopy became a mother lode of species, impossible to ignore.

The very notion of those nameless species, tucked away within dark sylvan folds, strikes a mystic chord in many people, lovers of arthropods or not. Lewis Carroll evokes this nicely:

> "What sort of insects do you rejoice in, where *you* come from?" the Gnat inquired.
>
> "I don't *rejoice* in insects at all," Alice explained, "because I'm rather afraid of them—at least the large kinds. But I can tell you the names of some of them."
>
> "Of course they answer to their names?" the Gnat remarked carelessly.
>
> "I never knew them to do it."

PAGE 98: As a caterpillar, New Guinea's *Ornithoptera priamus* birdwing butterfly feeds upon noxious vines of the genus *Aristolochia*.

PAGE 99: *Oecophylla* weaver ants, one of the most pugnacious species of Old World ants, repair a rent in one of their canopy leaf nests by pulling the leaves back in place.

"What's the use of their having names," the Gnat said, "if they won't answer to them?"

"No use to *them*," said Alice; "but it's useful to the people that name them, I suppose. If not, why do things have names at all?"

"I can't say," the Gnat replied. "Further on, in the wood down there, they've got no names. . . ."

The formal naming of a new species, be it a butterfly or a bird, is a purely utilitarian matter in science. It is the quest for their favorite organisms that provides the greatest thrill for most taxonomists.

ENTOMOLOGISTS ON A RAMPAGE: THE MATHEMATICS OF DIVERSITY

In Peru, Terry Erwin woke me and other assisting scientists before sunrise so we could avoid the breezes that stir rainforest canopies as temperatures rise. We tramped over soggy dead leaves through dank rainforest understory to where we had stretched plastic sheets beneath a tree.

Terry carried a contraption called a fogger, the quintessential tool for gathering canopy insect specimens. It resembled a chain saw from which a tube projected instead of a saw. With the pull of a cord, it gave out a chain-saw-like roar. White fog poured from the tube and drifted obligingly into the canopy. Terry walked beneath the tree directing the fog into every part of the tree's crown, the odor of insecticide trailing him. The first casualties fell within minutes. Soon falling bodies pattered regularly on the sheets. For forty-five minutes, insects showered down as researchers swept them from the sheets into alcohol.

The biodegradable insecticide kills arthropods but leaves vertebrates like birds and mammals unscathed. Terry finds that even the insect populations return to their previous levels in weeks, as insects recolonize a fogged tree from the surrounding forest.

I had reached this spot in Pacaya-Samiria National Reserve by houseboat, traveling the Amazon and its tributaries with Terry's group. River dolphins flashed pink backs at us and subsided into mud-brown or inky black water. We dozed in hammocks, eating piranhas lured with fruit on a hook. We savored whiskey, sunsets, and good conversation. And each day we gathered at tables amid a constant babble of Latin names, to sort that morning's insect catch.

Terry made his initial 1982 calculations of arthropod diversity using this same method of fogging and sorting specimens. He found a huge number of species in crowns of a certain Panamanian tree. He assumed a portion of those to be habitat specialists on that tree; that is, rare species likely to occur nowhere else. Given that there may be fifty thousand rainforest tree species, and assuming that each harbors an equal portion of distinctive arthropod species, he came up with thirty million arthropod species worldwide.

In the Amazon basin, Terry has now studied forests of different soil types and forests experiencing different degrees of flooding. Each tree community harbors a disparate insect community. Because tree communities vary so much in rainforests, this is further evidence of an enormous number of canopy insect species.

Some experts dispute Terry's calculations. With the little we know today, other scientists have predicted species numbers ranging from five million to eighty million. But, unquestionably, Terry's calculations have compelled biologists to reflect on species diversity and to recognize that most previous global estimates have been seriously low.

Insects are merely the supreme example of tropical diversity. Most organisms achieve their highest diversity here—and notably in what William Beebe thought of as earth's eighth continent, the rainforest canopy. This unparalleled wealth is not a law of nature. In a 1992 paper, Michael Mares shows that in South America, drylands support higher mammal diversity than rainforests do. Still, mammal diversity in the rainforest is extraordinary by temperate-zone

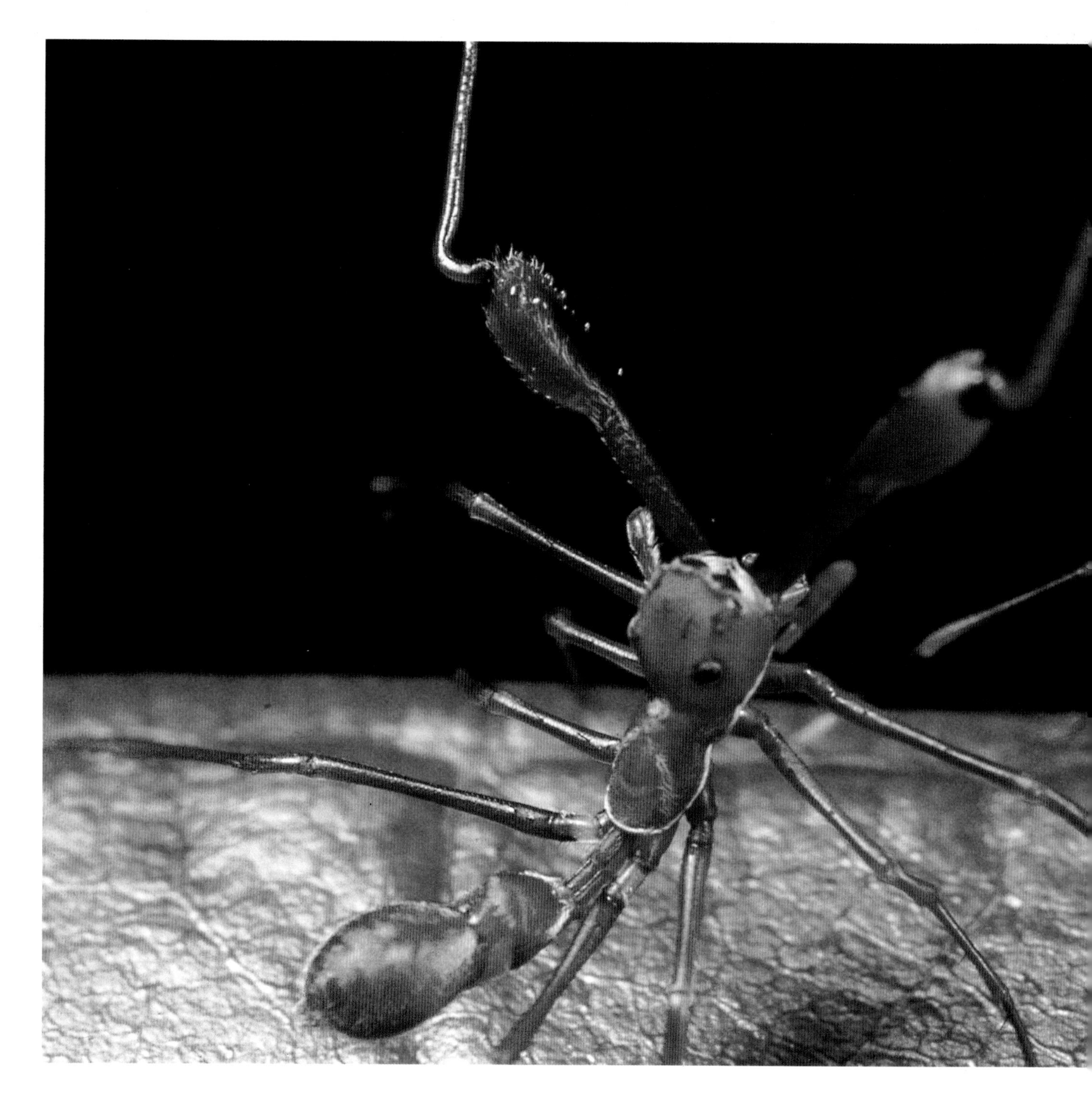

In the Sri Lankan canopy, *Myrmarachne plataleoides* jumping spiders unsheathe fangs and wield them like swords in a duel. The victor rears up like a boxer in the ring as the loser backs off. Spiders are one of the many invertebrate groups other than insects that are well represented in the treetops.

standards, and most other terrestrial groups do appear to reach peak diversity in tropical rainforests.

Insect Living-Space in the Canopy

Terry Erwin's research team during my stay included experts on tiger beetles, leaf beetles, ground beetles, and ants. Though Terry is best known now for his scholarly writings on general rainforest biodiversity, his

particular expertise is on beetles. Beetles are the largest and most diverse of the insect groups, and the gem-like beauty of many species makes them popular with collectors.

I'm amazed at how many people (even those professing a passion for beetles) seem to appreciate beetles not as living things but as objets d'art to be pinned unthinkingly. This view might change if aficionados met one of Africa's Goliath beetles or America's Hercules beetles. Some of these giants weigh as much as a songbird. In flight they assault the ears like a jet at close range. They show superb adaptations to the canopy, scuttling over trees on long powerful legs. When picked up they grip tenaciously. It may take two people to unwrap all six double-clawed legs of one Hercules beetle from another person's arm.

During my first trip to the tropics at age seventeen, I kept a male of one of the giant beetles. He ate a hefty portion of a banana every day, then kept me awake at night with his heavy breathing. That beetle never stopped reminding me that he was alive.

Yet canopy beetles, like most canopy insects, average a tenth of an inch (two and a half millimeters) in length. Most of Terry's efforts focus on one family of smallish beetles, popularly called ground beetles. Ironically, tropical ground beetles flourish in trees. "I'm trying to look at the canopy through the eyes of these two- to three-millimeter-long beetles," he says. With this in mind, in addition to fogging he's started to scramble about in the canopy to find out where they live and how they survive.

Terry has fogged single trees that have yielded hundreds of thousands of individual beetles and other insects from a thousand or more species. This is all the more impressive given that most of these species have a larval (growth) phase and an adult (reproductive) phase so radically different that each phase acts with total autonomy, in effect almost doubling the number of species in groups that spend their whole lives in the canopy. How can so many creatures find a place to live in one tree?

Trees provide the majority of living space in a forest: the leaf-surface area of a tropical tree may be tenfold the ground area below. The area of bark and of the upholstery of arboreal plants found in each tier of skyscraping rainforest canopy adds further to this enormous surface. Compared with other terrestrial habitats, rainforests offer a much greater area for supporting animal life than a map would indicate.

This bonanza of space is especially useful to insects, most of which are so small that gravity is largely irrelevant. In this regard it usually matters little whether an insect selects a vertical or horizontal surface, or for that matter whether it is on a tree or the ground. All else being equal, we would expect insects to select the canopy—and particularly the rainforest canopy, with its multistoried strata—merely because trees provide them with elbowroom. Indeed, although we know nothing about most canopy insects, we suspect that not every species in the canopy will be a full-time resident there. Some, which Oxford ecologist Sir Richard Southwood calls "tourists," may be just passing through.

Herbivores and Diversity

But all else is not equal, for the canopy is the center of forest production, holding the bulk of plants. Naively, we would expect, then, that there would be all sorts of opportunities for canopy animals. But to test Terry's views on the number of arthropod species, we need to know how the arthropods of the rainforest canopy (and especially the insects there) associate with particular tree species or assemblages of trees. There's little to go on: the sky is now the limit for canopy studies.

If animal diversity relates to the wealth of trees and tree communities—which is an assumption of Terry's projections—the diversity of arthropods (and presumably other organisms) should plummet when tree diversity is low. For this reason, comparisons of forests rich in tree species with monodominant rainforests (those with few tree species, like *Gilbertiodendron* forests of the Congo and *Mora* forests in Trinidad) will become focal points of future studies. Observations at ground level suggest that some forests with low tree diversity do have a scarcity of animals. Among the handful of tree types in Borneo's kerangas forests, the spectacular background symphony of insects, monkeys, birds, and frogs peculiar to Asia is all but replaced by a dead silence punctuated by the lone calls of cicadas.

Most of the arguments about diversity are based on scientific ideas about species interactions. Herbivory, for example, is a type of relationship between animals and plants often mentioned by Terry and others in the biodiversity numbers game.

Although leaf-eating vertebrates like sloths and howler monkeys consume many kinds of herbage, they do so at a cost. Because the durable foliage of many tropical plants accumulates a plethora of defensive substances, these voracious "generalist" herbivores have to stomach many unpleasant chemicals during their lives. It takes a convoluted digestive tract to detoxify all those leaves. Sloths need one hundred hours to digest one stomachful. Most generalists shift constantly between plant species to avoid cumulative overdoses of any one toxin. Also, leaves are so energy-poor that warm-blooded herbivores must be bulky in order to process sufficient food for their energy needs: no leaf-eating canopy mouse exists. (Other factors work against small canopy mammals: see Chapter 7.)

While some herbivorous insects are generalists, many of them avoid stomachaches by specializing on one or a few plants. Rainforest plants often generate fresh leaves throughout the year, making specialization highly practicable. Classic cases of specialization include the New World *Heliconius* butterflies that select certain vine species on which to lay eggs: the hatchling caterpillars stay put. They can handle the chemical arsenal of that plant and grow quickly there.

Many plants counter the onslaught of insect and vertebrate herbivores with a vengeance. Every plant is a unique battlefield of deterrents—tannins, alkaloids, and other substances—that often repel all but a few crafty enemies. A quarter of our prescription drugs are chemicals such as these, extracted from the living pharmacy within the tissues of rainforest plants.

The entire surface of Australia's gympie tree is laced with toxins that are acutely irritating. This nasty brew keeps most herbivores—and people—at bay. Yet a select army of insects has adapted to a gympie diet. The sap of the rubber tree dries into a sticky latex that glues up most would-be insect assailants. And so on: other tropical plants wield chemical weapons that block protein absorption; attack the nervous system; cause heart attacks; shut down DNA activity; or suppress the molting of insects so that as they grow they are crushed within their own unsplit exoskeletons.

Generally, the more long-lived the plant is, the more noxious it is. Climax trees live long enough to be discovered by myriad insect herbivores. Many climax species keep all but a few specialists at bay by growing inedible, toxic, or nutrient-poor tissues. Short-lived pioneers expend less on defense and more on growth. Though they can suffer high rates of leaf damage from a glut of generalist herbivores, it is easy for them to replace foliage in resource-rich light gaps.

The durability of individual leaves also matters. Many epiphytes have life-long foliage that is tough and inedible. The leaves of understory trees often appear to last longer than those of canopy trees. We might predict they would better defend themselves and thus, like epiphytes, stay relatively unscathed.

Herbivores can be finicky even on their preferred plant. Different herbivore species may select different parts of the plant, in effect dividing it among them. Many favor new leaves, which are easier to chew and more nutritious than old ones. This may explain why some trees flush many leaves at once, instead of replacing leaves a few at a time. By mimicking the spring bud-breaks of temperate plants, the leaves have a chance to harden before herbivores can eat them all.

To most herbivores, trees and their flora are islands in the sky, some hospitable, most not. Specialist herbivores add to the plants' difficulties by decimating the young of the host species. Finding easy pickings near parent trees, they ravage tender seeds and saplings and so contribute to the rarity of the host "islands." Only those young plants lucky enough to be dispersed far from others of their species escape the onslaught of herbivores and have a chance to grow to maturity (see Chapter 2). For this reason the life cycles of many rainforest insects presumably include long-distance journeys to colonize the widely scattered surviving plants. Certain tropical butterflies and moths have been known to migrate in dense flocks of millions, at least in some cases in search of food plants for their larvae. The Africa explorer Henry Morton Stanley witnessed one example in the Ituri forest:

> We saw a cloud of moths sailing up river, which reached from the water's face to the topmost height of the forest, say 180 feet, so dense, that before it overtook us we thought it was a fog, or, as was scarcely possible, a thick fall of lavender-coloured snow.

Fogging by Terry Erwin and his colleague Brian Farrell confirms that herbivorous beetles often occur on just one or a few types of trees, and so, like their host-plant species, they tend to be scattered sparsely in rainforests. Predatory beetles, in contrast, tend to range throughout a particular rainforest type but seldom cross between forest types (as defined by soil or seasonal flooding). These findings suggest that in choosing a habitat, predatory species respond to the architecture of the canopy, while herbivorous ones attune more to the chemistry of specific plants.

In short, rainforest canopies provide the greatest banquet of trees and other plants for herbivore specialization and so create opportunities for many herbivore species. They similarly provide bonanzas for other plant-eaters like pollinators and fruit-consumers (see Chapters 9 and 10). With so many species eating plants, parasites and predators can flock to the canopy metropolis.

Parasites: A Short Story

Of course, parasites and occasional predators like pythons are problems even for people. I tend toward good health in the tropics, but like all roving biologists, I have my stories.

Once, when I was in Trinidad as a student, the top of my head began to sting, but friends weeding through my hair saw nothing but a bump. After several weeks, I sensed something amiss on a walk to my office at the Museum of Comparative Zoology. Reaching up, I felt a soft object squirming from my scalp. I collapsed to my knees. At that moment I recalled a film in which a person ruptures open and a large alien emerges; could this be the probing finger of some greater horror within?

Rationality returned: it could only be a botfly. I pulled for several seconds before a rubbery white maggot the size of the last segment of my finger fell to my feet. Emptied of its fly cargo, the swelling on my head collapsed. Dizzy, and repulsed by this being that had consumed my flesh, I nonetheless recollected that I was an entomologist, and was thus obliged to pickle it.

Clasping the maggot in my hands, I dashed to the museum. I was on my way to a bottle of alcohol when one of my professors called out to me. He guided me into his office to eagerly relate new developments in our research project. Hands interlocked in apparent serenity on my lap, I realized I was not psychologically ready to show anyone this creature born of Moffett. All I could think of was whether my wriggling captive could chomp through my palms with the ease with which it had emerged from my head. I empathized with the monkeys parasitized by botflies I had seen in the canopy.

Despite my personal trepidation, I knew that botflies are innocuous unless they attack in exceptional numbers. And like many of the untold species of insects, protozoa, worms, and other parasites that victimize the tropical wildlife of the ground and in the canopy, their life history is fascinating.

Big and clumsy, the adult botfly avoids being swatted by its vertebrate quarry by laying eggs on mosquitoes. When the mosquito lands on a warm body to feed, out hatches a tiny maggot, which bores promptly through the skin. Feeding *ad libitum,* it breathes through a snorkel that it pokes through a small hole in the host's body. It avoids retribution from the host by jabbing its spines into the flesh when the least bit disturbed. Upon reaching full size—in forty or fifty days—it emerges from the host's body through its breathing hole, burrows into the soil, and transforms into an adult fly.

Magnified thirty-five times, a mite retreats to a natural chamber next to a leaf's midrib. When creatures visible at this high level of resolution are considered, the number and diversity of organisms living in one tree are almost beyond fathoming.

Ten minutes passed before I found a quiet spot and dumped the maggot in alcohol, where it sits today. If I had been more courteous perhaps I'd have let it transform into an adult. Even now this is one insect I've never told my former professor about.

Soil Creatures Move Up

Soil animals are the most unexpected and least understood canopy tenants. How they get around is unknown. Perhaps some negotiate directly from one crown to another, whereas others move from tree to tree by way of the ground. (I envisage this anthropomorphically as a journey requiring a worm's version of an intrepid explorer's mentality.)

Is this soil fauna unique? Focusing on the soil inhabitants of the cloud forest at Monteverde, Costa Rica, Jack Longino and Nalini Nadkarni found that species differ between ground and canopy but typically belong to the same or similar groups. Moreover, groups considered overwhelmingly terrestrial in other ecosystems—scorpions, earthworms, ground beetles, and so forth—manage nevertheless to invade soil mats in the canopy.

Did soil inhabitants evolve in the canopy or on the ground? It's hard to say, but flooding in the Amazon basin literally drives terrestrial invertebrates into the trees, perhaps accelerating adaptations to canopy life. Michael Goulding, an expert on flooded forests, describes this mass exodus:

> At the beginning of the rainy season . . . soil arthropods begin to migrate upwards to the trunk and canopy layers with spiders, millipedes and centipedes being especially common. Most arthropod groups appear to migrate before the actual inundation starts. . . . [Others], however, wait it out, and only leave the forest floor when it is flooded. Sow bugs (tiny crustaceans) and small spiders are among these adamant groups. Spiders especially, but also predaceous ants, form a veritable gauntlet that upward-moving, flood-fleeing invertebrates must run.

Some insects turn their food-plant's toxins to their advantage. They introduce the toxins into their bodies, making themselves inedible. These species often advertise their unpalatability with gaudy pigments and virulent hairs (TOP, Panama). Many edible insects employ camouflage, the opposite defense. Here a bark-mimicking jumping spider waits in ambush for a leaf-mimicking insect (RIGHT, Sri Lanka). Others mimic something dangerous. One moth has the eyes and legs of a jumping spider etched in its wings, perhaps confusing spiders that try to stalk it from behind (BOTTOM, New Guinea).

Michael says that some predatory arthropods presage the receding floodwaters, descending weeks early to near water level and ambushing creatures that climb down as the water drains.

No major new life forms have yet come to light in the canopy. Because the force of gravity is of little consequence to small, almost weightless creatures, perhaps it's to be expected that broad groups of animals seldom restrict themselves to the treetops.

A Beetle's-Eye View

The organization of the canopy insect community is incredibly subtle. Imagine the viewpoint of an average-sized beetle flying in a high canopy stratum, for whom lights and shadows represent pools of heat and cool, of dryness and moisture. A lone bobbing leaf within the clustered foliage signals a wisp of air that threatens to blow us off course—one of many currents funneling through cloistered passages in the vegetation.

Maneuvering through the latticework of foliage, we join the complex air traffic of the canopy. Minute aphids and thrips float by like aerial plankton. Below us a helicopter damselfly with a four-inch, ultrathin body leisurely beats diaphanous wings emblazoned with purple stains. It wobbles in the air before a spiraling web, daintily plucks out the spider, munches on its abdomen, and discards the rest.

Reaching a twig, we linger a moment, our wings flickering a hundred or so times per second, wing covers standing out stiffly above them. Propelled forward through an innate response to canopy structure, we continue from twig to branch, soaring deftly an inch or so above its lush green fleece of epiphytes. Our eyes convey the crudest of information. Yet our antennae now pick up a veritable soup of odors: singular organic traces from each plant that we pass; pungent whiffs of the humus beneath the plants. As we approach the tree trunk, the flora on the branch grows denser and richer.

Soaring in an arc over a bromeliad, we suddenly hit a swirl of a certain scent that sends us into a downward spin. Receptors on our antennae follow the chemical gradient until we home in on a space between two dagger-like bromeliad leaves.

When we land, the air feels cool and stagnant. A bead of water sits squatly next to us on the waxy leaf like a great glass globe, its surface etched with a thin filigree of bacteria. Against the leaf sits a millimeter-thick film of moisture-saturated air loaded with carbon dioxide. Tiny springtails and oribatid mites creep about at our feet, living and dying fully immersed in this film.

Skirting omnipresent canopy ants, we scramble to the odor's source within the earthy mass at the base of the leaves. Pushing into sopping, mangled decay, we alarm microscopic rotifers (with whirling cilia) and roundworms that push spastically through the soil. Two strawberry-red mites, the size of the period at the end of this sentence, climb off our back, where they had hitched a ride, and disappear into the dross. Probing eagerly with our antennae, we uncover a bed of mold. We pull out wads from the copious fungus and savor the thin hyphal strands. Mites of another species—one that permanently resides on our body—dash forward to drink the exudate on our face as we eat.

This story, I hope, provides an inkling of a basic truth: virtually every nuance of the canopy, including shifts in smell, texture, taste, temperature, and humidity too fine for human senses, registers with beetles and other arthropods. Selecting from a menu of possible canopy environments, some species may be fairly indiscriminate, but many, like our beetle, will be very picky. To the majority of the world's fauna, conditions can vary significantly over distances barely detectable to the human eye.

The canopy resembles a Russian nested doll: opening it, we find unforeseen layers, one after another. To borrow a term from mathematics, ecosystems exist in "nested sets." Each tree houses a community; each member of this community in turn houses its own community. Organisms migrate between communities, but every community acts in part independently, with its own special dynamics. And the communities are constantly changing. Using an ultraviolet light trap raised and lowered on a rope, biologist Henk Wolda has found remarkable (and inexplicable) species fluctuations in insects collected each night for fourteen years in a Panamanian canopy. Employing varied approaches in the field, ecologists (notably Nigel Stork from the British Museum and Yves Basset from the Bishop Museum) recently have begun to address basic questions about the structure and dynamics of these tropical canopy communities.

The Bromeliad Community

It so happens that bromeliads contain an exceptionally multifarious community of insects and other species. Explorer-naturalist Thomas Belt described the bromeliads he saw at every turn in Nicaragua as sitting "perched up on small branches like birds." In fact, bromeliads so dominate the Latin American canopy that the reservoirs they hold between their upper leaves form a landscape of fragmented swamps prized by arboreal species. Bromeliads are such stiff and spiny plants that this wildlife is normally hidden from us.

Like any community, that of the bromeliad encompasses its share of fierce predators. Dragonflies and helicopter damselflies have wingless immature forms called nymphs that grow up in bromeliad tanks. These kill prey with extendable jaws that they fling forward from under their heads, as do those species that live in rivers and lakes. A Jamaican canopy crab has soft-bodied larvae that damselfly nymphs find very tasty. The mother crab goes to great lengths to protect her minute offspring, clearing everything from the water in her bromeliad's tank before laying her eggs. She stands guard for a couple of months as her young develop, provisioning them with snails and millipede prey, and crushing trespassing nymphs and other predators.

Some vertebrates such as poison dart frogs likewise use bromeliad tanks as homes or nurseries. In the case of the frogs this is all the more remarkable because they are ordinarily terrestrial. Their tadpoles hatch from eggs laid in the ground litter and squirm onto their brilliantly colored parent's back for transport to the trees. Lacking the climbing toepads and agility of tree frogs, they find the trip upward arduous. Depending on species, either the father or the mother accomplishes this task.

In the most extreme example of parenthood, *Dendrobates pumilio* from Panama, the job falls on the mother. Searching the canopy, she deposits each tadpole in its own bromeliad tank. She withdraws from any site already occupied by a tadpole, which signals its presence by shaking its tail as she approaches. Scattering her young among plants assures that not all will be eaten by damselfly nymphs or crabs. Remarkably, the frog remembers where she put her young. Every few days she makes her laborious way back up the trees, visiting each tadpole in turn and depositing unfertilized eggs in its pool. Because her tadpoles deplete the available food, her eggs are an essential dietary supplement.

Such residents are valuable to bromeliads. As animals defecate or die they add to the compost that forms a bromeliad's source of nutrients. Some bromeliads are deemed to be partially carnivorous because animals drown and decay in their basins. True carnivorous plants (bladderworts) sometimes float in the tanks, trapping aquatic prey inside tiny bladders.

War and Peace in the Insect's Social World

A friend, listening to my tales of insects, once asked me why insect lives are filled with so much drama. I responded that perhaps, with so many rivals and enemies, small, short-lived creatures experience life at a greater tempo. Whether or not this is so, I believe there is a distinction between the research frontiers for vertebrates and for insects. Although we may learn new details about them, most vertebrates that we think of as charismatic—such as lions, giraffes, and bears—have been well known for centuries. By contrast, we will find radically new, drama-filled stories about previously unknown insects far into the future.

The social lives of canopy insects show extraordinary promise in this regard. Before climbing gear existed, naturalists armed with binoculars could not hope to see most canopy insects, let alone learn anything about them. An exception, tropical fireflies, deeply impressed many writers. Consider the words of Sir John Bowring, who wrote about them in 1857 after a visit to Thailand:

> How can I pass the fire-flies in silence? They glance like shooting stars, but brighter and lovelier through the air, as soon as the sun is set. Their light is intense, and beautiful in colour as it is glittering in splendour—now shining, anon extinguished. They have their favorite trees, round which they sport in countless multitudes, and produce a magnificent and living illumination: their light blazes and is extinguished by a common sympathy. At one moment every leaf and branch appears decorated with diamond-like fire; and soon there is darkness, to be again succeeded by flashes from innumerable lamps which whirl about in agitation. If stars be the poetry of heaven, earth has nothing more poetic than the tropical fire-fly.

Still, even such massive displays (sometimes with millions of fireflies drawing together to illuminate whole banks of trees) remain poorly understood. It may take someone willing to climb among the insects for a closer look to fathom their behavior.

The activities of social insects (ants, wasps, bees, and termites) in the trees are as spectacular as the mating aggregations of fireflies. Certain wasps and bees, protective of their nests, rank with the most menacing of treetop residents. Canopy termites, though less overtly hazardous, make me just as nervous. They turn branches into death traps, poised to snap under a slight weight.

Most termites infest dead wood, so climbers must take care that their branches are healthy—something that can be hard to determine from the ground. Foraging termites are seldom visible, being concealed inside galleries made of chewed wood. Yet the basket-like nests of *Nasutitermes*, the most conspicuous tropical-canopy termite, stick out at a distance like giant cankers on tree limbs, not auspicious omens for tree-climbing biologists.

Many tree-dwelling termites forage more on the ground than in trees. Perhaps it's safer below. There is a children's cartoon in which one character stands on a branch and saws it from the wrong side: whenever termites similarly eat the branch they stand on, part of the colony's labor force may end up plummeting to the earth. Many tropical trees are hollow; these interiors may be more suitable as feeding sites.

Ruler of the Tropical Rainforest: The Ant

From northern latitudes to the equator, ants become more and more dominant, ever increasing their legions of individuals and species.

That's especially true in the canopy. In the American tropical lowlands, nearly half of the insects taken by fogging are ants. While these ants sort to fewer species than do beetles, flies, moths,

Venezuela's predatory *Daceton armigerum* ants are strictly arboreal. TOP: Workers convulsively close trap-like mandibles on prey. MIDDLE: The ants retrieve a caterpillar while Chloropid flies dive-bomb them to sneak a taste of the food. BOTTOM: One ant lifts a back leg like a dog until another arrives to groom it there. Curiously, the ultimate ground predators—the army ants—have never taken up permanent canopy residence, although on occasion a few species raid in serried columns high in the trees.

or, for that matter, many other arthropods, Terry Erwin has collected as many as sixty ant species in one tree—a third of the number of ant species occurring across all of Europe.

Indeed, had I felt skepticism about the prominence of ants in the tropical canopy, my brush with the ant garden during my climb with Terry would have taught me a lesson. I can imagine the similar panic of pioneer canopy biologist Elliott McClure in

the Malay Peninsula. In 1960, while assembling a ladder during his first ascent of a rainforest tree, he encountered throngs of ants above ninety feet. Tales "of parachutists stranded in treetops only to be eaten by vicious ants" came to mind. Fortunately, the species he saw posed no threat to humans.

Ants, then, are a tropical success story. Yet one mystery is the universal absence of ants above elevations of about 7,500 feet in dense forests the world over, recounted here by William L. Brown, Jr., of Cornell, the grand master of tropical ants:

> This fact never ceases to surprise me as I start hopefully into lush forest (after considerable trouble to get there!) at say, 2300 to 2500 meters in the Colombian Andes, in the Nilgiri Hills of southern India, or in the Ankaratra of Madagascar. Even at 2100 meters in most tropical mountain forests, ants are exceedingly scarce, and in any one locality are represented by very few species. Yet at much higher altitudes (of 3500 or even 4000 meters) on treeless slopes of the Andes or the Himalayas, ants may be locally abundant.

In part to investigate this trend, Nigel Stork of the British Museum fogged insects at varying altitudes in an Indonesian rainforest. As altitude increased, ants declined, whereas mites, spiders, and beetles became more abundant. Bill Brown suspects that beyond a certain elevation, insufficient solar heat may penetrate the trees to allow ants to forage or their larvae to grow. Only when crowns are well spaced, letting lots of light energy through the roof of leaves to ant nesting sites in the branches and the ground layer beneath, do any ants survive at high elevations.

Another peculiarity of a few ants and some other arthropods is that ground- or understory-dwelling species in lowland rainforests sometimes move higher into the canopy with increasing elevation (although in many cases the individuals at different elevations may actually belong to closely related species distinguishable only to the sharp eye of the expert). This trend, which has also been suggested for certain epiphytes, may be related to the insufficiency of sunlight and heat near the ground, and the absence of arboreal competitors (like the nefarious weaver ants of the Old World or *Azteca* ants of the New World) that keep those species out of the trees at lower elevations.

TREETOP HOMES FOR ANTS

I made the acquaintance of the arboreal ant *Daceton armigerum* in 1977 while exploring Venezuela's Orinoco basin with two colleagues. Armored and spiny, *Daceton* are unmistakable: lobes at the back of their heads house adductor muscles that power their snap-action mandibles. A procession of lumbering workers descended one tree, but shied away from earth. A worker's feet touched the soil for a few seconds at most before she scurried back to solace on the trunk or foliage.

The nest was in a hollow twenty feet up in a small tree. In those days I was unable to ascend to the ant's world, so I chopped down their nest. Back at the hotel, the three of us transferred the ants to containers for shipment to the States. The whole business became awkward as ants escaped. Each one of 2,342 workers could grip anything—be it the nest, bathroom tile, or ceiling—as if her feet had been glued in place. I like to think we got them all, but doubtless strays startled hotel staff for days afterward.

Some ants nest terrestrially and forage both on the ground and in the trees, giving them a toehold on both environments. Yet there seems to be no shortage of treetop abodes for fully arboreal species. A few, such as *Daceton*, select recesses in wood. Flat "turtle" ants like hollow twigs. Certain dust-speck-sized ants prefer fissures in bark. Others lodge in epiphyte root mats.

Alternatively, an arboreal species may construct nests from scratch. These ants may use carton (chewed plant material), as do species that grow ant gardens. Or, like weaver ants, they may use live foliage; weaver ants shuttle their larvae back and forth to bind leaves together with larval silk.

Certain trees, epiphytes, and vines provide accommodations for ants—private chambers for them to colonize. Some plants cater to specific ants with living arrangements that only a few species can utilize; others are less particular. In return, the ants keep herbivores away, saving the plant the trouble of making energy-expending toxins. The ants also prune vines and epiphytes from their host. Instead of heaving refuse to earth, some ants dump decaying matter on special sites within their nest cavity where the host plant can absorb its nutrients directly.

The relationships between ants and plants have been studied for a century. As a high school student I enjoyed descriptions of *Cecropia*, an important pioneer tree of the American tropics. *Cecropia* trees house *Azteca* ants within their trunk joints. The ants feed almost exclusively on glistening white drops exuded from the leaf bases. Worker ants keep the tree clear of foreign plants and animals even in the area around the tree base. Like the *Camponotus* ants tending ant gardens, *Azteca* workers don't just pick on creatures their own size. If you so much as tap a *Cecropia* tree, ants will boil out of holes in its trunk and seize your finger tenaciously.

The ants that construct gardens provide places for the plants to grow instead of the plants furnishing space for them. Actually, when the large *Camponotus* ants in the ant garden attacked me in Peru, I noticed another species: smaller workers of *Crematogaster parabiotica*, shy ants that rushed about on my skin but did not bite. The *Crematogaster* and *Camponotus* often live together in ant gardens and even share foraging trails. They construct treetop carton nests up to several feet wide. Apparently, the workers collect seeds for food, some of which survive and sprout in the carton. Loosely speaking, the ants "grow" the epiphytes for their seeds: cactuses, bromeliads, figs, philodendrons, anthuriums, and orchids unfurl leaves from the carton, creating a bounteous miniature garden.

It's easy to assume that the plants require ants for their survival, for these particular species seldom occur elsewhere, and the ants protect nest and garden with zeal. But Diane Davidson of the University of Utah points out that the plants in ant gardens—like plants cultivated by human farmers—may not have a choice in the matter. The ants could be so thorough at snapping up their seeds that the plants may have little opportunity to germinate anywhere else.

In all these cases, both the plants and the ants benefit so profoundly that ecologists have treated their interactions as one of three major classes of mutualisms between plants and animals, on a par with pollination and fruit dispersal (humans and their crop plants might be viewed as a fourth class). Why *ants* as guardians and never, say, aggressive mammals or wasps? Because only ant workers will relentlessly scour a limited territory (like a plant's surface) in sufficient numbers to weed out enemies, large and small.

A similar mutualism exists between plants and mites. Many plants produce microscopic chambers on their leaves that mites use as protective retreats. We learned only recently that the mites consume fungi and creatures that could eat the plant—notably plant enemies too small for ants to handle. Unlike associations between ants and plants, those between mites and plants are widespread in cool climes as well as in the tropics. Resident ant societies need year-round living quarters, whereas mites can aggregate rapidly in the spring as soon as leaves open. The long-overlooked mite "bodyguards" demonstrate how easily we can miss vital interactions at micro-scales.

Treetop Ant Territories

Room for ants to nest may be plentiful in the canopy, but space for ants to forage probably isn't. In tropical Africa and Asia, hundreds of thousands of workers populate each colony of *Oecophylla* weaver ants. These patrol several trees, driving out competitors of the same and most other species. The workers lay claim to every piece of this territory by marking it with an odor peculiar to their colony. The ants use multiple communication systems, each one effective for a different foraging or defense situation. Foreign ants cause an immediate reaction. Alarmed by chemical signals, weaver ants converge on intruders, seize them, pinion them, and tear them asunder. The ants are so aggressive that adjacent colonies develop a "no-ant's land" between territories on the ground and in the trees, which no ant dares enter. Unlike *Daceton*, weaver ants have no qualms about coming to ground level, and the tree trunks are vigorously defended as likely breaching points for territorial invasion.

Up to two hundred weaver-ant leaf nests are dispersed throughout a colony's territory. Many hold a reserve of workers that can be drawn on for defense in that part of the territory, quickening the response to local emergencies. Only one nest contains the colony's crucial reproductive individual, the queen.

Ant-garden ants, *Azteca* ants, weaver ants, and others defend themselves too perniciously to be forgotten. Yet while ants dominate rainforests in aggregate, the majority of the thousands of canopy-ant species, like most ants elsewhere, easily go unnoticed. They creep out of view, run from trouble, or blend with the environment. Typically, they survive within the territories of whatever dominant ant species happens to overlook or ignore them.

And what convoluted territories canopy ants have! Ants do not deal with the entire canopy volume, just the expanse on which they walk. For them, a tree is a contorted surface. Like outer space bent by gravity, the everyday geometry of our world does not hold. Let an ant walk in one direction, and she soon returns to her starting point (she circled a trunk or branch); let her make a ninety-degree turn at any point, and she may never find her way home. Getting to and from a canopy nest can be a tricky business. Even more so than their terrestrial counterparts, arboreal ants rely on "scent" trails or other cues to orient themselves.

Costa Ricans call *Paraponera clavata* ants "bullets" because of their unforgettable stings. After finding cumbersome prey in the canopy, a bullet ant deposits a chemical trail to the nest at the tree base that guides assisting ants to the site. Reportedly, several hundred workers can be recruited to deliver valuable food. A potential canopy nightmare would be to confront such a swarm while on a rope, unable to flee. Each worker is an inch long, fearless, and built like a tank.

Because alternative routes exist among the interdigitating branches and vines of the canopy, trails to different places often cross. Bullet ants can tell apart trails laid by different nestmates, avoiding traffic confusion in the trees.

Weaver ants create shortcuts through the canopy labyrinth by bridging gaps between twigs. The ants interlink bodies, hanging for an hour or more in living chains that other workers walk over.

THE ROLE OF ANTS IN THE CANOPY

Big ants, little ants, mean ants, timid ants, what can so many canopy ants do for a living? More or less what ants do on the ground. Many scavenge for dead insects; others, like weaver ants and bullet ants, prey on live ones as well.

Few ants are totally carnivorous. In the canopy many species tend scale insects, mealybugs, or hopper insects the way ants tend aphids in temperate areas. These insects suck plant sap and, after absorbing some nutrients, excrete the excess into the mouths of eager ants. Whereas in themselves sapsuckers are harmful to plants, the attendant ants may benefit plants by devouring herbivores like caterpillars.

Working in Asia I noticed blemishes on the leaves of many dipterocarp trees. Later, I was amused to find an article on the "biology of green spots." As it turns out, the spots exude sugars that attract ants.

Such "extrafloral nectaries" are widespread among plants, both among those that provide housing for ants and those that do not. Ants feeding at nectaries will often guard them, and so incidentally keep the plant's enemies away.

Not all ants in the canopy are beneficial to plants. One of the few groups of ants totally dependent on vegetation are the leafcutters of tropical America. The ants raise symbiotic fungi for food in their subterranean nests. As fodder for the fungi, workers slice dime-sized pieces from leaves. Holding these aloft like banners, the ants stream to the nest in long cavalcades.

Principal herbivores of New World rainforests and scourges of tropical farmers, leafcutting ants (*Atta*) don't actually consume leaves, but raise a special fungus upon them. Fungus gardening is accomplished by an assembly line of worker sizes. Ants with heads about two millimeters wide cut and retrieve most leaf fragments. In the nest, crushing of the leaves and care of the spongy gardens are performed by successively smaller worker size classes.

Leafcutting ants select leaves with care. They prefer vegetation that is easy to cut; new leaves are ideal. However, the gaudy pigments (anthocyanins) in flushes of young leaves are toxic to fungi. This toxicity not only inhibits fungal infection in delicate, fresh leaves, it also keeps the fungus growers away. The ants go after older leaves, or younger ones without defenses.

Most ant species patrol circumscribed areas, divvying up the canopy into territories in much the way that humans partition the earth. They can alter the plants and animals living within their territorial borders, promoting the survival of some (like scale insects) but killing others (like herbivores). Thus, in most tropical forests, the rich mosaic of ant species in the tree crowns, which overlays the rich mosaic of the tree species themselves, can profoundly influence biodiversity by enriching the variety of living spaces available within the trees.

CHAPTER SEVEN

FURRED AND FEATHERED ON THE TOP OF THE WORLD

The spectacled bear spied a bromeliad in a branch fork above her and began to climb up the tree after it. I climbed an adjacent tree for a better look. From there I watched the two-hundred-pound bear tear the bromeliad loose and drag it to the ground. She and her cub consumed each thick leaf base as humans might eat an artichoke. Then she lumbered up another tree.

Suddenly my tree began to shake: the cub was climbing up to check me out. He was larger than I had estimated. As I anxiously shook my foot in his direction, mother let out a roar. I looked around. How can a person in a tree run or hide from an angry canopy bear? Fortunately the cub was as alarmed by the roar as I. He shimmied down my tree and bounded off with his mother into the forest.

A bird may fly over the canopy's surface, but neither a bear, nor an ant, nor any species simply ambles across it. Since empty space is ubiquitous, charting a safe and expedient course through the treetops is an ordeal for flightless vertebrates. All but the smallest or most languid of them must pass from tree to tree and from level to level routinely, often dozens of times a day.

One tactic is to find paths linking trees. No one has mapped all the potential routes in a rainforest but it's likely that most possibilities are either precarious or circuitous; very few may be trustworthy and direct. This became clear to ecologist Pierre Charles-Dominique in Cameroon. He captured poky nocturnal primates called pottos by affixing traps to sturdy poles secured between trees. Pottos turned out to be an easy catch: they soon abandoned natural routes in favor of the ideal pathway that a pole offered.

Some climbing species develop an intimate memory of the network of routes available to them. Longtime observers know that much of the traffic flows over a few choice paths in the canopy. How do these roadways arise? How long do they last? As yet no one can say for sure.

The Canopy Highway

Where epiphytes abound on the eastern foothills of the Andes, some aerial routes are as neatly demarcated as a hiking trail. I think about these beaten paths during interludes in the foliate chaos of such a canopy. What seems a likely sequence for the genesis of a canopy trail unfolds in my mind . . .

A column of leafcutter ants extends up a great kapok tree, where the workers fan out to harvest soft, young leaves. Several spider monkeys swinging hand over hand through the tree pause to observe the ants. Faced with the airy fathoms at the crown edge, the monkeys grasp branches tightly with tails and feet and stretch toward the neighboring *Anacardium* tree. They pull themselves over once they reach a convenient branch. One mother spans the abyss while her tiny youngster scrambles over her body, a living bridge, to the other tree.

Later that day, as the spider monkeys return to their sleeping tree, they pass the ants still at work on their own trail.

Some scouts from one ant column that extends twenty-five feet along a stout, epiphyte-laden bough venture onto a broad liana that drapes over the branch and loops across the abyss to the *Anacardium*. There they find a wealth of pristine vegetation. As they return, bearing freshly cut leaf banners, each ant rubs the tip of her abdomen along the new trail, depositing a chemical called a pheromone that stimulates nestmates to follow the same path.

As traffic on the trail increases, some ants prune back plants that get in their way. Others gnaw at the trail's surface until it is smooth and easier to negotiate. Beneath arcades of spiky bromeliads

PAGE 116: Living on Andean slopes, the spectacled bear (*Tremarctos ornatus*) is the most arboreal of any bear, even constructing canopy sleeping platforms of broken branches. Succulent bromeliad hearts are its predominant food (Colombia).

PAGE 117: Emerald toucanets (*Aulacorhynchus prasinus*) forage at all canopy levels in small groups of their own kind (Costa Rica). Mated pairs of other tropical birds often join flocks of ten or more species to scare up insects or plunder fruiting trees.

along the branch, the cleared trail develops into a narrow earthy brown streak running through a cushion of moss.

Two days later the ants abandon both crowns, having exhausted the fresh leaves of the kapok and *Anacardium* trees. That evening an arboreal mouse chances upon the deserted trail. Taking advantage of its cleared surface, he pushes under saw-edged bromeliad leaves and scoots along the liana. Stuffing his cheeks with epiphyte seeds, he scampers back to eat in his moss-lined nest in the kapok. He adds the trail to his regular forays, chiseling at bromeliads until he clears a mouse-wide tunnel through the plants.

Whereas bird diversity declines after rainforest is fragmented, small-mammal diversity rises. A surfeit of species like mouse opossums (*Marmosa*) may be a sign of a habitat in decline. An orange-red pelt is peculiar to nocturnal canopy species (Costa Rica).

FACING PAGE: The green palm viper *Bothrops lateralis* detects the direction and distance of warm-blooded prey with an infrared-sensitive pit located in front of each eye (Colombia).

One night a green tree viper lies in wait on the bough. The mouse saunters along, oblivious. Suddenly the viper strikes. The mouse leaps. The snake grazes his side, deflecting him over the brink. Down he plunges, limbs spread, twenty, forty, sixty feet, leaves slapping him on the way. He hits an understory branch and bounces. He gets up from the branch, grooms, and hurries home.

Not surprisingly, the mouse avoids the trail for the next couple of days. When he warily returns, the snake is gone.

A month afterward, a newly weaned woolly opossum in the *Anacardium* wanders onto the trail. Squeezing past the epiphytes, she finds a comfortable nook in the kapok, clears out the frass, and moves in. Thereafter she forages in both trees for fruit and insects. The *Anacardium* becomes her prime hunting ground, however. Like many tropical trees, the *Anacardium* has rotted and decayed in its core, forming a hollow cylinder replete not only with night-active bats but with opossum delicacies—scorpions, centipedes, cockroaches, and termites scuttling over the caked bat feces. Each night, the opossum enters the towering grotto through a branch stump in the crown and pursues live prey amid fungi and dried insect husks.

Five months go by. The mouse has been eaten by an ocelot, but the opossum's continual attention to the trail keeps it clear. A silky anteater occasionally finds the pathway convenient and eventually another mouse incorporates it into her schedule. Since the once-sheltered trail has widened, this mouse is more jittery than her predecessor, bolting over exposed segments in a blur.

All the while, spider monkeys have been passing through the kapok on their way to fruiting trees. An independent youngster deviates from his parents' course and discovers the path along the branch and liana. Within a week the group shifts its routine. Now they can avoid the chancy span from the kapok to the *Anacardium*, striding down the trail instead, tails held out straight for stability. The bough's upper face becomes trodden into a channel. In some places matted plants extend from either side like hair along a part.

Generations of vertebrates come and go along the major canopy thoroughfare. In addition to the nightly voyages of a prehensile-tailed porcupine or a kinkajou, the daylight hours see the sporadic passage of a squirrel, collared anteater, howler monkey troop or band of coatis looking like long-nosed raccoons. Once in a great while a tayra—resembling an oversized chocolate-brown mink with a toothy grimace—fords the bridge, gulping down mice or dozing lizards. When the kapok comes into bloom, a family of night monkeys exploits the trail at dusk to eat the lavish white blossoms.

Gravity and Body Size in Climbing Animals

Gravity is an important issue in rainforest biology not only because so many vertebrates—amphibians, reptiles, and mammals—climb into trees, but because so many of the climbers are *large*.

South American spectacled bears and African gorillas, both of which top four hundred pounds, represent the size limit for temporary canopy incursions. While climbing, both species treat every move with utmost care. To get to the next tree, they must descend and detour over the ground. That's their only secure route.

Animals of smaller size must find more direct solutions for commuting between trees. Whether one is a canopy worm (Chapter 6), a wandering monstera plant (Chapter 5), or a monkey, as long as one has the capacity to handle slender tree limbs, heading earthward is a waste of time and effort.

Orangutans, the largest full-time tree dwellers, must avoid any mishaps in the trees because a fall would be deadly. Adult orangutans pluck fruit or tear up moss mats for termite morsels while suspended between branches, but two or more of their gripping hands and feet are always firmly planted.

When flustered, green iguanas weighing several pounds plunge suicidally from perches eighty or more feet high. The lizards often bask over rivers, where the water absorbs the impact of their fall—though broken bones are commonplace among them (Panama).

Straddling trees where stout branches grow cheek by jowl, they step gingerly from one to the next. Larger individuals seem to know that their bulk is a liability, and two-hundred-pound males may take roundabout courses over the ground to avoid chasms that they could have bridged in their youth.

Leaping is one of two basic styles that climbing animals can use to catapult themselves through trees: they either leap with their legs or brachiate with their arms. Leapers include many species, such as macaques, leaf monkeys, and squirrels, that spring from atop horizontal branches, and a few, such as tarsiers, that hop between vertical boles. Brachiators, including spider monkeys, woolly monkeys, and gibbons, swing arm over arm under the branches; it is easier for heavier animals such as these to hang from slender branches than to balance on top of them.

Because most branches grow upward toward the light, animals leaping from tree to tree often land on twigs and leaves. As Ivan Sanderson describes, flimsy branch tips can sustain this impact because the monkey spreads its weight among its extremities:

> The monkey then lands, not on the top of a branch, as is popularly supposed, but on the side of a mass of leaves and smaller twigs, with its arms and legs spread-eagle fashion. It grasps the foliage in an all-embracing hug and then scrambles to safety.

Crossing trees in this manner is done piecemeal. After landing in foliage an animal has to wait for the branch to stop oscillating before it scrambles to a firmer surface to make its next leap.

Only a few species of primates brachiate. Those that do it best are the Asian gibbons, which vault through space up to thirty feet at a time. I was first introduced to gibbons in 1983 by Mahidol University biologist Warren Brockelman at Khao Yai National Park in Thailand. The startlingly white lar gibbon was so far up in the canopy pinnacles it looked like a minute cloud. For me its agility was a revelation: after looking down at us, it shot along from limb to limb as if it were a bird in flight, each time flinging its arms overhead to grasp firm branches. Suspended below swaying branches, brachiators seem wondrously graceful in trees.

The most gymnastic leapers and brachiators are experts not only at bridging trees but at bypassing the meandering course of a tree's architecture, almost making a beeline in the canopy. Leapers bound across branches near the core of the tree, slowing down at the flimsy margins of its crown. Brachiators do better on flimsier branches, skirting the center of the crown where branches are too broad for them to grasp from below (they can resort to leaping or to traveling on top of branches by foot if they have to).

In Uganda's Kibale Forest, Colin and Lauren Chapman of Harvard University introduced me to the locomotor skills of primates like the black-and-white colobus, redtail and blue guenons, gray-cheeked mangabey, and chimpanzee. The Chapmans track the animals from the ground below. When I described the advantages of rope-climbing for canopy research, Colin was dubious. "Maybe you need ropes to study canopy ants, but once monkeys have habituated to us, we can collect all the data we need using plain old binoculars from the ground." Then he mused, "Floating above them in a little balloon—now *that* would be nice."

Some monkeys studied by the Chapmans seem to throw caution to the wind. One day I enjoyed the rhythmic crashing of red colobus as a troop lined up to leap in turn from one perch onto another. Dashing on ahead, one daredevil sprang to a branch that broke beneath its weight. The monkey fell sixty feet, still attached to the limb, and landed with a resounding boom. I rushed forward expecting to find a mangled monkey, but the lucky creature merely raised its head, bared its teeth at me, and bounded up the tree.

When a bough does break, rather than panic the animal often stays calm. Back in the 1960s, Elliot McClure observed this reaction in a siamang, a Malaysian relative of the gibbon, from his 120-foot-tall platform (as Andrew W. Mitchell writes):

> Perhaps the siamang was not paying enough attention to where he was going, but as he swung onto a dead limb it snapped, dropping him into the green abyss below. He uttered no sound, showed no concern, made no frantic struggle—just watched as the canopy passed until a limb offered itself, then reached out and took it in his grasp, swinging smoothly back on course in his original direction.

This apparent lackadaisical attitude is deceiving: for an animal of several pounds, a fall can injure or kill. Once as Yum, an Indonesian assistant to primatologist John Mitani, watched a family of Bornean agile gibbons 120 feet overhead, a two-year-old lost his grip while playing with his parents. The youngster plummeted to the ground and died. According to studies from the 1950s, more than half the population of wild gibbons has healed fractures from past falls.

For mice or small opossums, the canopy is more forgiving. No tree is so high that dropping from it can kill or even cripple them. Creatures in this size range seem to take falling in stride, as naturalist Thomas Belt witnessed:

> I was once standing near a large tree, the trunk of which rose fully fifty feet before it threw off a branch, when a green *Anolis* [lizard] dropped past my face to the ground, followed by a long green snake that had been pursuing it amongst the foliage above, and had not hesitated to precipitate itself after its prey. The lizard alighted on its feet and hurried away, the snake fell like a coiled-up watch-spring, and opened out directly to continue the pursuit. . . .

Even some petite monkeys show a remarkable indifference to falling. On sighting a hawk, South American tamarins will tumble in all directions; from the ground, it may appear to be raining monkeys.

Animals smaller than a mouse may be able to cling effortlessly to trunks and the undersides of branches, but, like the mouse, they still must contend with the effects of gravity. An organism must be considerably smaller than the average insect before the dynamics of falling change substantially: spores, pollen, cysts, and the minutest insects fit that description. These lilliputians are almost as likely to go up as to go down. When cast into space, they drift mostly at the mercy of canopy eddies whether or not they bear wings. Thrust on a similar course by breezes, larger insects join the lilliputians and the nutrient-rich aerial dust that incessantly flows through the trees.

As a graduate student at Stanford University, Roman Dial—since 1992 a professor at Alaska Pacific University—set out to determine how important airborne drifts of insects are to arboreal *Anolis* lizards. In a forest on Puerto Rico's rainy Luquillo Mountains he hung sticky drums to catch aerial insects. Roman found that the lizards mainly lived where the most insects intercepted the canopy. When he ousted the lizards from these trees, insects on them increased in numbers. In addition, the leaves became riddled with holes. He deduced that the lizards had been keeping herbivores in check. The consequences of losing the lizards had cascaded through the ecosystem.

Communication in the Canopy

I met Cathy Langtimm at Monteverde, Costa Rica, in 1990. Cathy, at the time a doctoral student at the University of Florida, Gainesville, conducted part of her canopy studies at night. Late one afternoon I came along as she climbed a tree, shouldering a backpack filled with sound-recording gear. We settled in the crown as the sunset faded to a faint orange and the uppermost stratum of foliage above transformed into enigmatic inky blotches against the slate-gray heavens. As darkness obliterated even these vague figures, Cathy waited patiently to record the two-note calls of arboreal mice with a shotgun mike. The loss of sight amplified the small, quick sounds of animals and the sway of the bone-hard branch under me.

Among canopy vertebrates, large diurnal monkeys and birds receive the lion's share of attention. Cathy has broken new ground in the study of such often-neglected creatures as mice and opossums. In her work with John Endler of the University of California, Santa Barbara, Cathy has found that the tone of an animal's pelt matches the brightness in its part of the rainforest, so that to predators active at the same time of day it appears to blend with the background. To accomplish this, most nocturnal mammals in the canopy are orange-red, whereas those on the ground tend to be dark brown or gray.

Communication with one's own kind is a challenge in rainforests. During the day, visual signals work well for canopytop species like the gaudy macaws. But for residents of the forest interior, members of one's kind are likely to be hidden from view behind curtains of foliage. In this region, visual displays tend to be less flamboyant; indeed, many rainforest species are somberly attired. Whether active by day or night, residents of these lower strata generally rely much more on acoustic than on visual communication.

How far sound carries depends on foliage and temperature, which in turn depend on the altitude in the trees and the time of day. Low frequencies

Atop a Costa Rican crown at dusk, Cathy Langtimm listens for *Reithrodontomys gracilis,* an arboreal mouse whose two-note calls may communicate location to prospective mates in the maze of branches.

TOP: Despite its awkwardness on the ground, a three-toed sloth in Panama descends a tree once a week. It digs a pit at the base of the trunk with its stubby tail, defecates, urinates, covers the pit with litter, and heads up the tree again. BOTTOM: Several species of beetles and mites, as well as pyralid moths that flutter in and out of its pelt as if it were an old carpet, leave the sloth's body at this time to lay eggs in its dung. In addition to these residents the sloth's hair has grooves that foster the growth of two species of blue-green algae.

FACING PAGE: Macaques lack prehensile tails, so stretching across a gap is not an option for them. Although young macaques eagerly play at branch tips, in most species older ones can't afford to be careless in trees (India).

are especially effective within sound-reflecting foliage, and they travel far along the sun-warmed layer of air just above the canopy. Territorial roars of New World howler monkeys or African black-and-white colobus—among the loudest rainforest residents—can be broadcast more than a mile. High-pitched squeaks dissipate quickly in the cluttered air, but are practical in the understory when animals live in close proximity and are small (thus not physically able to generate low frequencies).

The intricacy of treetop vocalizations varies too. Complex patterns that incorporate many tones tend to advertise the whereabouts of their creators, which may be good for attracting mates or warning rivals but which may also draw the attention of predators. To keep enemies from homing in on the call, many rainforest animals produce a simple, narrow frequency range of sound and begin and end the call gradually. An alternative is to use a more discreet mode of communication, such as plumes of odor drifting through the air. Undetected by humans, each branch and leaf in the canopy probably has been scent-marked by some crawling insect or mammal, to indicate a choice route, to attract mates, or to warn others of territorial rights.

Adaptations for Climbing and Leaping

The anatomy of two- to twenty-pound canopy residents often deviates markedly from the anatomy of ground dwellers. Arboreal species may have long, flexible spines, mobile-jointed limbs, and feet that grip, often with pads and ridges that provide extra traction. Ground animals and occasional canopy visitors have stiff tails that primarily aid in balance; in some tree inhabitants, like the spider monkey, the tail can grasp and has a sensitive pad like a fingertip at its end. In extreme cases, the muscular tail weighs as much as one of the animal's legs and can support its whole body.

Tree-dwelling species in this same size range show a diversity in structural adaptation and carriage associated with their methods of locomotion. Animals that walk atop branches often have short limbs or they move in a crouch: a low center of gravity keeps them from toppling over. To deal with branch spacing, they also have a more flexible pattern of footfalls than mammals on the ground. In leapers the legs are longer than the arms; brachiators have long arms for lengthy swings and their elongate fingers quickly hook and release branches. Thumbs, which might snag vegetation as a brachiator swings along, may be reduced in size, particularly in spider monkeys.

The most specialized canopy species are about as inept out of trees as fish out of water. Explorer Charles Waterton gave his assessment of the three-toed sloth in 1825: "This singular animal is destined by nature to be produced, to live and to die in the trees; and to do justice to him, naturalists must examine him in this, his upper element." A sloth hangs beneath branches by its long claws. On the ground most only can drag themselves.

A spectacled bear visits the canopy often, but still looks much like any other bear. Perhaps life for it is a compromise: unable to stay aloft permanently, a bear has to get about both on the ground and in the trees. Its proficiency in trees is largely the result of behavioral adjustments rather than morphological ones.

Canopy adaptations are also subtle among the smallest vertebrates. The feet of a canopy mouse are likely to be wider and shorter than those of a ground species, its big toes may be opposable, and its tail somewhat prehensile. Extreme anatomical peculiarities are absent in part because a mouse isn't threatened by gravity. Crawling to a tree's fringe may not require any skills beyond what it would need for climbing a shrub on the ground; its claws serve the mouse just fine. Furthermore, the crest of a large branch may seem precarious to humans but is as level to a mouse as are most places on the ground.

A mouse-sized species can also travel up and down trees easily by getting a grip on a tree trunk with its claws. For larger animals, ascending and descending a trunk is a struggle, though wrapping long arms around the trunk can help. Wide trunks can be impossible for the heaviest climbers. Chimpanzees prefer to use stalwart lianas as ropes to pass through regions where options are limited, as in the aisles of trunks between canopy strata.

Claws are not the only way in which small vertebrates cling. Salamanders, only three species of which climb trees in the temperate latitudes, are important canopy residents in the American tropics. As many as thirty individuals have been found in a single bromeliad, where they are among the top resident carnivores. At a maximum length of two inches, these animals are little canopy trapeze artists, with long limbs and prehensile tails. They stick to plants by surface tension, squashing their moist, wide bodies and expanded feet against the leaves

and squirming their way about. The reduced pressure within the water film between salamander and substrate holds the two together as long as the water adheres well to both surfaces.

So effective are the toe pads of tree frogs and certain gecko, skink, and *Anolis* lizards that these animals can even cling to glass. It was once thought the pads worked by suction—negative air pressure. Now it is known that frogs rely on surface tension to adhere to foliage, just like salamanders do. Lizards, on the other hand, have dry skin, and so they rely on microscopic hairs or minute folds on their toe pads to adhere to substrates as smooth as leaves.

None of these mice, salamanders, lizards, or frogs would have difficulty on low vegetation. When small vertebrates confine themselves to treetops, it is seldom because anatomy restricts them there, but because of a preference in diet or breeding site, or to avoid competitors or predators on the ground.

Caution in the Canopy

Until the completion of a helicopter pad in 1992, Keyt Fischer's remote mountain site in Papua New Guinea was accessible only by a grueling two-day trek from Mianmin, the nearest village with an airstrip. Hiking up a slippery path at the site—seemingly no place in the New Guinea highlands is flat—we looked for the subjects of Keyt's dissertation studies at Harvard: cuscuses and tree kangaroos. At the 5,100-foot elevation, the ground and the slender trunks of the tall trees around us were spongy with moss.

Cuscuses and tree kangaroos are among the menagerie of canopy marsupials native to New Guinea and Australia. Neither seems a promising candidate for treetop life: kangaroos are slipshod climbers, and both kinds of mammals move sluggishly for arboreal animals of their size (though one tree kangaroo species can approach thirty-five pounds, most are less than twenty pounds, and cuscuses weigh five to fifteen pounds). Looking much like their more famous ground-dwelling counterparts, tree kangaroos differ in having shorter feet and somewhat longer arms. "Tree kangaroos climb trees the way repairmen climb telephone poles," Keyt told me. "The kangaroo wraps its arms around the trunk and digs the formidable claws on its feet into the bark like a pair of cleats, then inches up. It moves along branches balanced like a tightrope walker, but mishaps are common."

On distant continents, many species the size of most tree kangaroos move through the trees with grace and speed. Why are New Guinea's tree kangaroos so inept? "I suppose they don't need to do better," Keyt said. "They have little competition here—no native primates, civets, or squirrels."

Certain other species also climb more cautiously than their weight would seem to dictate. In the lethargy business, the 8- to 18-pound sloths of the Americas put cuscuses and kangaroos to shame: even switching trees is accomplished only once every day or two. A sloth's diet of foliage grants it no energy to spare (see Chapter 6). Metabolism reduced, it has to sunbathe to maintain its body temperature. The three-toed sloth must select a course in the trees with care; if it can't locate tender leaves and a place to sun, its digestion could slow down so drastically that it would starve on a full stomach.

Old World chameleons, some of which attain a foot in length excluding the tail, are also glacially slow for their size. They hold their leaf-shaped bodies above a branch on long legs, gripping with pincer-like feet. With their methodical surefootedness, chameleons proceed to the twigs without swaying or shaking them, as a similarly large animal

FACING PAGE: Keyt Fischer sallies into the crown of an enormous *Pandanus* tree in New Guinea. The increasing frailty of a tree's limbs as they bifurcate outward and a dearth of lianas tough enough to support a person's weight impede rope climbers from going out on the thin limbs where most of the action in the canopy ecosystem occurs.

with clawed feet might. By this means, they avoid alerting their quarry: flying and jumping insects or, in a few cases, birds.

Predators of fierce or quick prey face special difficulties in the trees. Unless blown by wind, branches transmit vibrations better than earth does. The thinner the branch or the bigger the predator, the worse the problem. Another difficulty is retaining a grip on both the branch and the struggling prey. The solution to both problems is to move unobtrusively and keep a firm hold on the tree—as the chameleon does. Some animals—arboreal opossums, primates such as the potto and lorus, and lizards such as New Guinea's emerald monitor—have grasping feet and a painfully sly approach to hunting. Yet chameleons outdo them all: in effect, they are always one step closer to their quarry thanks to a sticky-tipped tongue, which they use harpoon-style, that can extend as long as their body.

Arboreal snakes must show similar stealth. With few places to hide, many of their prey exploit sites that transmit vibrations distinctly: lizards, for example, often sleep on twigs. Yet tree snakes can spread their fantastically drawn-out bodies lightly across many branches, and so approach their target without shaking its perch. A prehensile tail allows a snake to lift its rigid forebody like a cantilever to reach the quarry, without even being on the same branch.

SIZE AND STRATA

The canopy's uppermost corridors hold a plethora of climbing mammals of moderately large size: monkeys, lemurs, cats, civets, porcupines, and so on. Surprisingly, smaller species, like rodents and opossums, are not as prevalent overall in high tropical canopies as they are in most habitats globally. What works against small mammals in large trees? Probably several factors.

Small mammals, being warm-blooded, suffer hardship in the exposed and turbulent environment of the uppermost trees. Because a small body has more surface area per unit of weight than a large one of similar shape, it gains or loses heat more swiftly. Thus, in the trees, where shelter from heat and cold is scarce and fluctuating conditions are hard even on epiphytes, a small mammal will have trouble maintaining its body temperature (the problem will not be so acute for "cold-blooded" reptiles and amphibians).

Small size makes it easy to scramble among twigs for insects, flowers, or fruit, but small mammals meet their match in large ones that have their own tactics for browsing among food-rich twigs. The weight of a gibbon hanging below a branch arches the terminal leaf spray down so that a volume of fruit-bearing foliage dangles in the ape's face. Walking or leaping species of a similar or even larger size access the outer twigs either by snapping off and retrieving the whole branch or by clutching stiff branches with grasping feet or tail and plucking food with their hands. Balancing carefully, six-pound Asian giant squirrels manage the same thing using only their claws.

Small climbing animals may reach twigs readily, but it is harder for them than for large climbing animals to cross the wide gaps from one crown to the next that typify the high canopy. A macaque or gibbon can hurl itself farther than a mouse can: it can achieve a better running start, and it can more effectively use a branch as a springboard; some monkeys bounce on a limb several times before jumping. Finally, the forward momentum of a small animal is seriously reduced by air friction against its relatively large surface area.

An inability to span open gaps in the upper canopy may be especially problematic for the many small mammals that supplement their insect diet with fruits and seeds. Trees with these foods can be sparse (see Chapter 9). Species that eat only insects are less affected because they may collect all the nourishment they require in a small area: in the

By hanging beneath branches, chimpanzees can wander to the twigs of figs and other trees to scrounge for food. Like most primates, chimpanzees have forward-directed eyes that provide excellent depth perception for climbing (Uganda).

upper canopy, a couple of lianas heading in the right directions might fulfill all their transit needs.

Rainforest trees are most widely spaced in parts of Africa, making navigation tricky wherever lianas are missing. Perhaps it is no accident that large climbing mammals are abundant in these locations. Rainforests such as Kibale in Uganda are dreams come true for primatologists. A dozen treetop primate species coexist, most of them large. But in this same reserve there are only two species of squirrels and few records of smaller climbing mammals.

Among climbing vertebrates, a body weight of about two to twenty pounds seems to be the ideal compromise for life in the canopy's upper echelons. Most species in this size range have less trouble bridging gaps than either smaller or larger animals do. Counter to most people's intuition, the higher the species of this "advantaged" size class climb, the safer and simpler the canopy environment may be for them: massive crowns of upper-canopy trees offer thick horizontal branches, and when lianas are present they must be rugged to withstand trees swaying in the winds.

Small species may be more at home in the woody grottoes of understory trees. These trees tend to be flimsy, their branches rising steeply toward the rarified light above. Most heavy animals shun such rickety supports for travel from place to place. Among the large residents, only the orangutan has an effective method for spanning delicate verdure: using its bulk to rock the whole crown of a pliant understory tree, it reaches across to the next crown and lumbers over!

Gliding and Parachuting

I met K. M. Chinnappa in 1981 during my brief stay at an Indian forest reserve called Nagara Hole. Chinnappa was a tall, lean forest ranger with a broad smile. I had heard he was the best animal tracker in the region, so I was ecstatic when he invited me to join him for a nighttime stroll in the "jungle."

As we went, Chinnappa sniffed the air. At one point he smelled a kill and we scanned the vicinity for a tiger but had no success in locating it. Another time we detected the familiar circus odor of elephants. Chinnappa provoked the hidden animal to trumpet by waving a flashlight erratically in its direction. We saw deer and Indian buffalo and the gaur, another enormous bovine. This was electrifying, but Chinnappa kept an eye on the canopy, insisting there was something living in the trees that I'd like even more.

At last we spotted two luminous eyes. Chinnappa told me to wait. Shrouded in the dark I heard him scramble barehanded amid a dense wreath of lianas until he was forty feet or more up a trunk. Goaded by his shaking, a rectangle the size of a house cat glided silently from the crown. It passed me overhead and continued about fifty feet to another tree, where it landed on the trunk with a squishy thump. While it shimmied up the tree, I identified it as a giant flying squirrel. When Chinnappa came down, his hands were swollen from the poisons of hairy caterpillars, but he was beaming. Eagerly, he asked me, "Did you see it? Did you?"

An unfamiliar glider can be an eye-opener. When a Dayak hunter brought a *Chrysopelea* snake to Robert Shelford at the Sarawak Museum in Borneo and claimed it had flown out of a tree, Shelford was skeptical. "So I took the snake up to the veranda of the Museum and threw it into the air," Shelford recounted in 1916,

> but I was disappointed to see it fall in writhing coils to the ground, which it hit with a distinct thud. Then I allowed the snake merely to . . . glide rapidly through my hands, straightening itself out and hollowing-in its ventral surface as it moved; this time it fell not in a direct line to the ground but at an angle, the body being kept rigid all the time. . . . There can be no doubt that the hinged ventral scales of these snakes, enabling them to draw the belly inwards, are a modification of structure rendering a parachute flight possible.

Gliding is one form of controlled falling. How does an animal control its fall? By maintaining the right configuration—limbs outstretched, fingers and toes spread, and belly flattened—the animal maximizes the undersurface it presents to the air. If the area of this surface is large relative to the animal's weight, air drag slows it down like an open parachute.

Among normally shaped or modestly flattened species, there are many that don't seem to control their fall: they simply turn head over heels and land hard. But species adapted to the vagaries of canopy life often assume an outstretched posture. Although some species manage a modest forward progress in the air, among species that do it on purpose this form of controlled fall—called parachuting—may be most suitable for escaping enemies or to get from canopy to ground, not for travel from tree to tree.

Parachuting can be a regular mode of transport. In 1979 two students working in Puerto Rico's Luquillo forests told Margaret Stewart, a professor at the State University of New York at Albany, that "coqui" frogs had rained down on them at five in the morning. It was several years before Margaret sorted out the coquis' unusual lives. The frogs hide near the ground on hot days, when the moist conditions are more desirable than those in the drier canopy. At dusk they scale the trees in large numbers over a fifteen-minute period, just after day-active birds retire, and just before nocturnal predators—such as tarantulas and whip scorpions—become active on the tree trunks. The frogs disperse in the damp nighttime canopy to stuff their bellies with the bounty of insects. Then they parachute to earth before dawn, saving time on the commute and avoiding most predators.

Only species whose anatomy presents a flying-carpet-shaped surface to the air, like the flying squirrel or snake, have an angle of descent so shallow (less than 45 degrees from horizontal) that we can legitimately apply the term *glide.* The species overcome the problems of small size and large surface area by using air friction and lift production to advantage in bridging gaps.

At four inches in length, Wallace's flying frog is sizable for a canopy frog, but so thin that bones jut beneath its chartreuse green skin. With flaps of skin along its limbs and between its toes adding further to its area, this Bornean frog can glide. Studies of artificial frogs in wind tunnels have shown that the body design of the flying frog gives it maneuverability in the air. Indeed, they swoop from tree to tree as a matter of habit, and even glide down to the wallows of pigs and tapirs on the ground, where they breed.

Among the most agile gliders are *Draco* lizards, the "flying dragons." *Draco* species are common in Southeast Asia but often go unnoticed because with their "wings" folded against their flanks they look like any other lizard. In fact, they can soar like butterflies forty or more feet between trees, spreading their black-and-orange wing membranes to more than half the length of their body. Looking across at the canopy from a hillside in Indonesia, herpetologist Karl Schmidt observed an exciting confrontation among *Draco:*

> A female with wings folded, clinging to the tree about twenty feet off the ground, was being courted by a head-bobbing male. As he moved head-downward toward her, his brilliant lemon yellow dewlap and orange wings were extended. This bobbing display had gone on for only a minute when a second male glided over from another tree and the two began to fight. They shook one another so vigorously that they fell off the trunk, separated in mid-air, and glided away to different trees.

Gliding and parachuting are practical among animals of a narrow size range: above a few pounds, an animal's velocity becomes perilously high and maneuverability difficult; below a few ounces, any breeze can push a glider off course.

Geographic Mysteries: Gliding and Prehensile Tails

Each discipline of biology has its quirky mysteries. For canopy biology, one is the disproportionate abundance of gliding animals in Southeast Asia, including Wallace's flying frog, flying snakes, flying dragons (*Draco*), two genera of flying geckos, flying lemurs, seven genera of flying squirrels, and flying opossums (in New Guinea). Africa has only a few spiny-tailed flying squirrels, while tropical America has no examples at all—even though temperate North America has gliding squirrels. Less proficient parachuting species like the coqui probably abound everywhere.

Scientists have attributed the profusion of gliders in Asia to the scarcity of lianas connecting trees there. The loftiness of Asian dipterocarp forests also provides gliders with the takeoff height they need to travel far; after landing, gliders scramble high into the new tree for their next skydive.

Equally puzzling is why the Americas have so many more prehensile-tailed animals—silky and collared anteaters, kinkajous, spider monkeys, woolly monkeys, howler monkeys, and many opossums, porcupines and salamanders—that all boast tails muscular enough to support their entire body. In Africa, only arboreal chameleons and scaly mammals called pangolins have such tails; Asia claims the weasel-like binturong, while Australia and New Guinea offer cuscuses and some sugar gliders. Prehensile-tailed snakes are worldwide.

The prevalence of prehensile tails in the New World has been attributed to the fragility of its lianas: African lianas may have to be tough to withstand abuse by elephants and other powerful ground mammals. In this view the addition of a fifth gripping limb is advantageous in New World animals because bridges between trees are unreliable (in Africa, many monkeys have no tail at all). Another attribute of the American tropics is the prevalence of palm trees whose fruits are desirable to many mammals. The palms shed lianas and have slippery leaves, and so prehensile tails may help hungry climbing animals reach their meals.

In the Fast Lane: Flight

With their enviable freedom to roam, birds and bats are among the definitive treetop animals. Their adroitness is inspiring. Consider an interaction that biologist Archie Carr witnessed in Honduras:

If you lie on your back and stare upward to where the treetops interlace high above, you may see pairs of bug-like hummingbirds zipping over and under the green canopy like dogfighting planes diving into and out of the cloud banks. One such pair may suddenly plunge downward in a breathless spiral, the hind bird following in machine-like detail every intricacy of the mad course of the pursued. They collide in mid-air in front of your face with what ought to be crushing force, and then fly separately away to sit on twigs and preen and await the fine new surge of anger that will tune their incredible little muscles for more joyous combat.

Bats and birds have circumvented the obstacles associated with small size in trees. Able to flit from branch to branch as well as to commute over vast distances, the most enterprising winged hunters search out rare delicacies and fly back each dawn to cozy roosts and nests. Yet walking or flying, they still have to contend with canopy architecture.

Flight has not been studied in rainforest settings for reasons that are manifest in Carr's description. Charting the plants and insects of the canopy is arduous enough; measuring the actions of a bird would be daunting. Nevertheless, the body plans of bats and birds can be rough indicators of where they fly.

Large species with wings spacious for their size often fly over the canopy, or alternatively (as with the argus pheasants of Asia and the curassows of tropical America) in the open space near the ground. Canopytop examples like hornbills and fruit bats need lots of elbow room, but they fly fast—a vital skill in the upper canopy, where most food is widely separated among giant tree crowns.

Small species with relatively small wings have a more maneuverable flight and find it easier to forage within the foliage; hummingbirds and certain bats can even hover. Some species that have moderate body weight and a midsize wing proportion adopt intermediate strategies; an example is the Sac-winged bat in Central America, which prowls for insects in the passageways between canopy strata.

Like climbing vertebrates, birds and bats show varied skills at clinging to or walking in the foliage to harvest food. Some birds, like Panama's striped treehunter, snatch insects from dead, rolled-up leaves wedged between branches—a tactic seldom reported in the temperate zones. Other birds, like woodpeckers and wood creepers, cling to trunks to glean insects from the bark. Still others forage among twigs, competing with the mammals that reach branch-tip food with gripping hands. Toucans and hornbills overcome the limitations of their cumbersome size by employing the bills as "arms" to pluck fruit from the twigs. Smaller tropical birds also have distinctly longer, narrower bills than temperate-zone birds. This configuration may help them snap their bills faster and thereby catch katydids and other fast tropical insects.

Moving Through the Rainforest: Disturbance and Structure

Climbing with Jay Malcolm in 1990, a year before he finished his Ph.D. at the University of Florida, Gainesville, was a humbling experience.

An assistant and I had hauled twenty pounds of my gear to a forest plot and spent an hour positioning a rope in a tree. Jay arrived as we finished. Watching curiously as I arranged clanking metal paraphernalia on my climbing harness, he took from his pocket a five-foot band of fabric and tied its ends together. Walking to a tree next to ours, he wedged both his feet halfway into the loop. With legs spread, he pulled the loop taut, then gripped onto each side of the trunk with the arches of his feet, wrapped his arms around the tree, and began to forge his way to the top.

Jay waited for me at a forked branch as I climbed my rope. It was hard to imagine that his loop served as well as all my accoutrements, yet Jay's loop is one of the oldest climbing tools, known to Amazonian Indians as a *pecohna*. To use one takes stamina and confines the climber to trees under sixteen inches in diameter. But the pecohna permitted Jay to achieve what few rope climbers can: the scaling of many new trees each day. Using this method, Jay had set up hundreds of mammal traps in the trees. By rigging a frame with a pulley system in each treetop, Jay could subsequently raise and lower his traps from the ground to check his catch.

Jay was a participant in the "Biological Dynamics of Forest Fragments" project based fifty miles north of Manaus, Brazil. Around us groves of differing size had been carved out of the wilderness and intervening tracts slashed and burned to create pastures.

"These forests have two mammal communities: an arboreal one, and a less diverse terrestrial one," Jay said. "Few species bridge both." Jay finds that in the forest patches the arboreal community tends to move lower into the understory than it does in continuous forests. Canopy birds show the same trend. To compensate for shortage of space, animals may be expanding their territories vertically. "This couldn't be the explanation for mammals," Jay told me. "The same pattern occurs along the edge of continuous forest, where there's no lack of room." Jay thinks that because the edges of forest patches and continuous forests suffer high rates of treefalls from wind exposure, their understories receive enough light to support the dense vegetational growth on which insects thrive. Canopy mammals and birds, most of which relish insects, may be following their stomachs into the understory.

Patchiness is a quality of deep forest, too, as where climax and pioneer trees intersperse. A canopy animal such as an opossum or monkey may not discriminate between such areas unless its size poses a problem in traversing frail pioneer branches. Canopy birds may more easily discern forest types, such as stages in forest regrowth. From a bird's-eye view, regenerating forest materializes as a pale canyon encircled by hills of darker green. Some birds specialize on pioneer glades, seldom straying far from them. Other bird species seem to move freely across the forest types.

Jay Malcolm uses only a loop of cord between his feet to climb trees and erect traps to survey mammals in a ten-hectare rainforest fragment near Manaus, Brazil. On the horizon stands a one hectare plot, likewise part of the "Biological Dynamics of Forest Fragments" project.

Thinking about such natural rainforest settings while walking back to base camp with Jay, I felt saddened. The charred expanses between forest plots were sizzlingly hot, ugly, and all too familiar to me from my travels. I longed for a return to the deep forest, to see monkeys pass by on their pristine canopy highways.

CHAPTER EIGHT

A FLORAL SYMPHONY

A pivotal player in the destiny of Southeast Asia's rainforests once resided in a clear film canister in my shirt pocket. At odd moments I would remove the canister and rotate it between my fingers. A stalk of wilted cream-colored flowers tumbled within, goading the resident—which resembled a dust mote—to emerge from hiding among the petals and scoot along the stalk. Later I tapped the creature from the canister onto the fifteen-story-tall dipterocarp tree from which I had plucked it. I lost sight of the dot within the congested greenery as swiftly as I would a needle tossed into a towering haystack. My speck was a thrips, classified on a par with beetles, butterflies, and wasps as a major subdivision of insect life.

Back in 1976, Simmathiri Appanah had no knowledge of thrips, except perhaps as a dim memory from an academic lecture. Appanah, then an eager graduate student at the University of Malaya in Kuala Lumpur, Malaysia, was exploring the pollination of certain rainforest dipterocarp trees mostly native to that country, a group of what foresters informally call Philippine mahoganies.

By that time, ecologists could no longer view plants as mere backdrops upon which animals sit; plant and animal lives, it was understood, intertwine. Tropical rainforests offered rich fodder for studies on pollination, seed dispersal, protection of plants by ants, and herbivory, but most work had been done with investigators' feet planted on the earth. The canopy tapestry, the epicenter of rainforest activity, remained largely uncharted territory.

Appanah was a canopy pioneer without realizing it. He concentrated all his energies on solving research puzzles. As with any pollination biologist, the first and most basic puzzle facing him was: how was pollen transferred from one flower to another? Back then, pollination in dipterocarps, the dominant tree family in Southeast Asian rainforests, was a mystery that had stumped scientists for decades.

Philippine mahoganies tower into the canopy. Appanah watched their flowers from a sling-like seat (a bosun's chair borrowed from its usual nautical application) hung from a twenty-two-yard-long aluminum pole, or boom, a device invented by Peter Ashton of Harvard. Its seesaw-like design allowed Appanah to reach any point in three-dimensional space with one assistant and minimal gear. Versatility and mobility were crucial, because the flowers grew at the tips of lofty twigs in trees scattered

PAGE 134: As Sri Lankan scientist Savi Gunatilleke watched her assistants strap a series of ladders up a *Shorea trapezifolia* dipterocarp tree, she confided, "I was too nervous to climb at first, but each morning when bee pollinators arrived, I could hear the whole tree buzz. After a time, I could no longer bear my husband's monologues from overhead about what they were doing." She laughed. "That prompted me to take the leap into the canopy . . . in a manner of speaking."

PAGE 135: A secondary hemiepiphyte in the New World genus *Columnea* has red-bordered leaves that draw in hummingbirds to the yellow fan of flowers beneath.

RIGHT: Simmathiri Appanah's assistant, Raffee, inventories the seeds of a *Shorea pauciflora* tree from a bosun's chair at the end of a maneuverable boom that permits him to reach the branch tips.

through the rainforest. Booms remain one of the least arduous, most economical methods of canopy access, and they barely leave even a scar on the boughs. It's surprising they haven't caught on with researchers elsewhere.

Recently an assistant of Appanah's demonstrated the boom for me at Pasoh forest in Malaysia. Tying a rope to the boom's midpoint, he winched the boom forty yards up with a block and tackle strung from a branch. Strapping himself into the bosun's chair that hung from one end of the boom, he raised himself into the canopy using a pulley system. On the ground, Appanah guided his assistant inside the crowns by maneuvering a rope that extended from the boom's opposite end.

As we watched the assistant far above us, Appanah

(now a scientist with Malaysia's Forest Research Institute) told me of his luck in 1976. Dipterocarp trees flower seldom and unpredictably. But within his first three months of work on the boom, flowering began—the largest bout of flowering in eight years. Appanah had to work quickly, because the dipterocarps bloom over a period of just two and a half months. He wouldn't have had another chance.

Appanah spent days squeezed in the creaking bosun's chair, sunlight gleaming off the receding quilt of leaves below and the occasional whoosh-whoosh of giant hornbills in flight close overhead. He began to worry. The sticky pollen was not blown away by the wind; surely, the heavy, sweet scent of the nodding flowers—as many as a million on one large tree, each a quarter-inch wide—should attract some animal to disperse it. Unfortunately no likely pollinators seemed to visit the flowers.

"I didn't find a thing coming to the flowers during the day," Appanah told me, his wide forehead furrowed at the memory. "But I knew the flowers didn't open until an hour before sunset." At night Appanah had been pestered slightly by nocturnal throngs of specks—insects smaller than gnats—but like any seasoned naturalist, he had accepted this as a minor occupational hazard. "Finally it dawned on me that something might be happening with these minute insects on the flowers at night. They were so small I'd been ignoring them until then. I didn't even know what they were right away." It was one thing to see floral visitors and identify them as several species of thrips, but quite another to prove that they were pollinators.

In the deep canopy night, the silvery glimmer of the moon on the tree crowns replaced the sun's harsh rays, the flutter of bats passing by overhead replaced the noisy hornbill wing-beats. Appanah watched the thrips closely. He pruned flowering branches and relentlessly continued his observations in the laboratory.

Step by step Appanah managed to fit the story of thrips together. A thrips' life turns out to be short and sweet. Females lay eggs on developing flower buds. In two days the young hatch and begin to rasp holes in the petals; six more days and they themselves are breeding. Adults live a week or two, walking between adjacent blooms or flying awkwardly on their feathery wings, all the while devouring oily pollen and, especially, the succulent petals.

"In the canopy emporium of poisonous foliage, the one hamburger joint is the petals," Appanah's colleague, Peter Ashton, told me. "Petals are seldom defended. It's not worth the trouble; they're too ephemeral." Thrips seem to take advantage of the easiest nutrition around.

A thrips in a dipterocarp tree flower (*Hopea nutans*), magnified thirty times. There are 4,500 species of thrips globally, very few of which are pollinators.

But, Appanah wondered, do the trees get anything in return for the damage to their flowers? He noticed that the oil on the pollen grains made them sticky enough to adhere to a thrips' body. When a thrips moves onto another flower it often lands on the stigma (the female reproductive surface) and preens, pushing off any hitchhiking micro-sized pollen, which then may fertilize the flower.

An element of the puzzle was still missing. Some plants can be fertilized by their own pollen, but this was not true of Appanah's trees. To be successful as dipterocarp pollinators, then, thrips must carry pollen not just from flower to flower, but from tree to tree. Did these feebly flying specks ever travel so far?

Pondering the thrips' movements, Appanah found that air currents were the key. Each batch of flowers lasts one night. In the morning, the bell-shaped rings of petals, with stamens attached, loosen and waft to earth, drifting as they go like small inverted parachutes. Adult thrips fly off as they fall, but wingless juvenile thrips stay with the petals to eat whatever tasty food remains on them or on the stamens attached to them. As night falls, new canopy flowers open on the same trees; a cloying scent pervades the air. Any newly matured thrips, pollen grains adhering to them, respond by rising in weak zigzagging flights, forming a haze. Once above the forest canopy, breezes may carry these pilgrims from one tree to another during this daily cycle.

It takes a lot of thrips to handle the immense number of blossoms. Several thrips species begin to flourish weeks prior to mass flowering, because many more of their offspring survive on the hospitable environment of dipterocarp flower buds. A female can lay thirty eggs during her brief life, so population levels swell rapidly. Densities climb so high that some thrips take to guarding a corner of their flower, driving off interlopers.

Few thrips survive the lean years between Philippine mahogany blooms. Survivors dwell on whatever flowers they can find, often those of pioneer species that breed out of sync with other trees. Like other thrips species, they seldom pollinate these flowers. They stay on the plants as inconsequential parasites, nibbling harmlessly on floral tissues.

Appanah's observations confirmed that tiny, wind-borne thrips indeed were the pollinators of the Philippine mahoganies. "Nobody conceived of such a thing for a tree a hundred and fifty feet in height," Appanah said as we crooked our necks looking up one of his trees. From these initial observations, reproduction in Asian rainforest trees has grown into one of the great sagas of canopy biology.

Sequential Flowering

In a tropical rainforest in South America or Africa, a variety of trees undoubtedly will be in flower at any given time. But in Southeast Asia, canopy flowers are normally scarce. I count myself lucky to have visited Appanah at the end of one of these flowerings, in July 1990. This was only the second mass flowering in Peninsular Malaysia since Appanah's first study on Philippine mahoganies in 1976.

In general, the forests of Southeast Asia spawn a floral riot like these once in three to ten years over a period of a few months. During such times, fallen petals of all shapes and sizes accumulate ankle-deep on the forest floor. Without a rope or boom it's hard to know which tree giants the petals come from: their reproductive frenzy is veiled from our view behind "dwarf" understory trees.

In the temperate zones and in monodominant forests in the tropics, mass flowerings occur, but they involve only one or two species that flower annually. The flowering orgies of Asia encompass a hundred or more tree species at a single, species-rich site. Often, nearly all the dipterocarps flower, and as many as nine out of ten of the canopy tree species join in as well.

Appanah studied six closely related species of Philippine mahoganies at Pasoh forest. He has found that during a mass flowering, they bloom in an order that is always the same. The thrips do not distinguish between the trees, which means that if all six had bloomed simultaneously, most pollen would have been wasted by thrips hauling it to the wrong tree species. Appanah concluded that

PAGES 140–141: *Vochysia ferruginea* trees bloom annually on the Osa Peninsula of Costa Rica. The half-inch yellow flowers are frequented principally by large bees, which must commute long distances between crowns. Bees are the most significant pollinators in tropical rainforests worldwide, with the possible exception of New Guinea's forests.

sequential flowering ensures that each tree species has all of the thrips to itself for a while.

During the actual flowering, both thrips and the trees dependent on them prosper. Because of some overlap in flowering periods, the thrips maintain their populations by transferring pollination activities from one tree species to the next.

Appanah compares Philippine mahogany trees to oaks and other temperate-zone trees that produce vast numbers of tiny flowers. Oaks dump their pollen into the wind. The pollen drifts to other trees by chance, without the assistance of animals. The wind stratagem is common in temperate zones, but, like the allergies wind-borne pollen exacerbates, it is almost unheard of in the tropics. Wind is indiscriminate about depositing its cargo and most tropical species are too highly scattered for the method to be effective. Rainforest plants usually rely on vigorous animals, such as bees or bats, to move pollen accurately over the great distances between neighboring plants of the same species. Yet Philippine mahoganies seem to be casting their reproductive fate to the wind because their thrips fly ineptly, drifting in the air currents. Still, their pollen has a better chance of reaching other trees than pollen carried by wind alone. Thrips may be weak fliers, but at least they can, with a bit of rowing, choose where they ultimately land when the wind chances to carry them near a flower. Also, like many dipterocarps, the Philippine mahoganies often occur in clumps, so the distances between trees need not be overwhelming.

Appanah has gleaned evidence that other participants in mass flowering have the same reproductive strategy as Philippine mahoganies, but with sequences of their own. In each sequence several tree species share a unique group of pollinators. Leafhoppers pollinate one series of dipterocarp trees; bugs and flies may pollinate others. Trees other than the Philippine mahoganies employ thrips as pollinators, but bait their own set of thrips species.

Depending on the species, bromeliads are pollinated by bats, insects, the wind, or birds such as hummingbirds (Costa Rica). In many bromeliads, the above-ground portion dies after flowering.

Each tree species avoids intermixing pollen with other trees that use the same pollinators by blooming only briefly during a mass flowering, waiting its turn to flower in a sequence. And by using different pollinators, each series of trees avoids intermixing pollen with the tree species in other series.

Leafhoppers, bugs, and flies that pollinate dipterocarps resemble thrips in ways that are crucial to mass-flowering trees. Able to subsist between blooms by eating foliage or plant sap, they apparently breed rapidly enough at the onset of flowering to handle the abundance of blooms.

A bout of mass flowering may extend over an area as small as one river valley or as large as Peninsular Malaysia plus northern Sumatra—a region the size of New England. All those chains of trees, all those throngs of pollinators, are interlinked in one enigmatic explosion of canopy procreation. No one knew what ignited this reproduction bomb until recently.

In 1986, Appanah teamed up with two colleagues to analyze the matter. "Consider the six Philippine mahogany species that live together," he said. "The first species in the sequence blooms ten days, the last one blooms twenty-five days. It's like a telescope—each tree in the sequence lengthens the time a bit more. When we do a regression on the telescope, we come to fictional species: we imagine trees with less than ten days of flowering that would fit in the same sequence, mathematically. Now we work back to an imaginary tree with zero days of flowering. Of course this can't happen, but that day, we argued, is the day the trigger was pulled to start the flowering relay."

When the scientists checked conditions prevalent on day zero for several mass-flowering events, they saw a pattern: nighttime low temperatures had dropped by at least two degrees centigrade below average. In the uniform climate of Southeast Asia, such a temperature shift is a rare event. Yet this modest gunshot launches the vast relay races of mass-flowering trees.

How did these grand pageants arise? Dipterocarps may dominate Southeast Asian rainforests today, but they probably came from farther to the north, from seasonal Asia. As do many trees that reside in seasonal regions, seasonal dipterocarps today use cool spells at the onset of the annual dry season as cues to prepare for reproduction. After their ancestors migrated toward the equator, the newcomers retained this behavior as a relic of their past. In uniform climates, they might have to wait years for a cool spell, but this means little to trees that presumably can outlast several centuries.

The key enigma of the rhythm of Asia's canopy flowering may be solved, but each tree species offers Appanah fresh riddles to drive his work forward. "Every time a dipterocarp flowers, my home goes nuts," he admitted. "Flowers in bottles, flowers in bags, flowers everywhere. These are the great moments of my life."

Payoffs for Pollinators

Why don't all pollinators show up at the same plants? No animal ever comes to a plant except for its own selfish reasons (and so birds probe epiphytes to drink or bathe in the water they hold, or to gather nest material or insects; insects eat leaves or hide among them). Similarly, no visitor fertilizes a flower without expecting a payoff. Plants tailor flowers to the selfish cravings of selected pollinators. We appreciate flowers for their beauty, but most floral visitors regard them as a meal of nectar or pollen laid out appetizingly in a decorative bowl of petals.

A Philippine mahogany flower seems downright unprepossessing. But it's made to satisfy thrips, not people. The flower supplies dinner: pollen and, in this case, the petals themselves. To lure thrips to this feast, there is no need for floral bulk or color; thrips have poor vision and don't need space. Odor matters most. The floral perfume entices thrips skyward each evening. Also, thrips don't settle down until they've discovered a crevice. The flower provides them with a compact meshwork of internal parts and plenty of hiding places among them. In contrast, a leafhopper's flowers are wide open and develop an engorged central projection (the style) from which the hoppers suck plant sap.

Some flowers circumvent the cost of food gifts like this by deception. Orchids are virtuosos of such fakery. My favorite is a species that uncannily resembles a female bee; naive males transfer pollen while attempting to copulate with the dummy flowers.

Other orchids and bees play another game. Male "orchid bees" pollinate 650 New World orchid species. They come for fragrances in the petals, which they mop up with their forelegs and shift to spongy tissue on their hind legs. The right mix of scents likely attracts females seeking partners to beams of forest light in which the males have gathered, buzzing riotously in groups of fifty or more, their metallic bodies glittering. The male bees can find the same perfumes in plants other than the orchids, but the orchids cannot reproduce without the bees.

Pollination Specialists

All the mass-flowering trees of Asia that scientists have examined so far provide food for their pollinators. Not all of them feed the small thrips-like insects we've discussed so far. Other trees that join the Asian flowering bandwagon lure pollinators of another sort: bats, birds, bees, butterflies, and moths. These animals manage to be full-time pollinators by shifting their foraging habits to wherever flowers are most abundant. For instance, carpenter bees that normally patrol treefalls and riversides—where the fast growth and brief life spans of trees mean that flowers are always present—switch to tree species deep in the forest at the onset of mass flowering.

Luring such pollinators is costly for the tree. Thrips need only nibble on a pollen grain, but full-time pollinators demand substantial meals. Bats are especially gluttonous. Pale bat flowers open at night, luring bats with pungently sweet odors and extravagant drafts of nectar or volumes of pollen. Some New World bats hover almost as gracefully as hummingbirds while they lap floral meals with long, agile tongues. Other bats cling awkwardly to flowers that grow stalks robust enough to bear their weight.

Birds, being (mostly) daytime animals, visit flowers that are distinct from those pollinated by thrips or nocturnal bats. The plants advertise their nectar gifts by blooming brilliant colors such as red, a color that works like a charm on birds. Hummingbirds (America's renowned ecological counterpart to Asia's sunbirds) investigate any red blotch within the canopy's visual imbroglio; even a red shirt excites them. Their flowers tend to be unscented, not because hummingbirds can't smell, but possibly because scents appealing to the birds may draw less effective visitors like bees. To prevent the entry of any visitors but hummingbirds, each blossom is tube- or cone-shaped. Nectar lies deep in the flower's throat, much like a sip of wine at the base of a slender glass. While a bird drinks, pollen rubs onto its beak or head. (Hummingbirds depart from flowers bearing

LEFT: Opening swiftly enough to attract the eye, flowers of *Pseudobombax septenatum* trees have stamens that give them a resemblance to shaving brushes, dusting attendant bats with pollen. Nectaries at the swollen base of each flower attract ant guards that drive floral thieves away.

BELOW: The flowers of balsa trees open near nightfall and are pollinated primarily by bats, although monkeys and birds such as yellow-rumped caciques arrive the next morning and rob them of any remaining nectar (Panama).

more than just pollen. Nectar-loving mites not much larger than the pollen grains hitchhike between flowers in the bird's nostrils.)

Because of the expense of doing business with such full-time pollinators, trees relying on them during mass flowering invest their reproduction in fewer, more extravagant flowers, rather than a profusion of small ones. The payoff for the trees is the distance some of these spunky pollinators can and will go between flowering plants: in some cases, miles. In this way, birds, bats, and bees are able to service plant species that have sparse populations.

Most full-time pollinators must have a continuous, even supply of food, year in, year out. In Asia, the paucity of food sources for them outside of mass-flowering episodes means that these animals are scarce. They also breed slowly compared with fecund thrips, and thus, despite their flexible foraging behavior, the number of trees that pollination specialists can handle on short notice is limited.

Over evolutionary time this quandary has resulted in a relationship between the diversity of dipterocarp trees and pollinator types. Those groups of mass-flowering trees reliant on slow-breeding pollinators often have few species, and only one or two can survive in any forest. Those reliant on fast-breeding insects, such as thrips, have many more species. Several of these species can coexist in a rainforest by forming flowering sequences serviced by exploding pollinator numbers.

ABOVE: To reach the flowers inside a ripening *Ficus brotryocarpa* fig, a fig wasp has to push her way between the scales circumscribing the minute hole at one end of the fig.

LEFT: Stationed on the fig, ever-resourceful weaver ants devour wasps as they attempt to penetrate its entrance (New Guinea). Figs and their wasps occur worldwide.

The thrips' strategy helps explain the species richness of Asia's rainforest trees. "Trees in a sequence do the same things," Appanah said. "They live on the same soils, grow to the same canopy levels. What are the differences between their niches that keep them distinct as species? As far as I can tell, just a couple of days' difference in flowering times, once every decade or so."

Figs and Fig Wasps: A Short Story

There is another group of trees that has proliferated wildly because of an unusual and effective pollination strategy. The more than nine hundred fig species include trees, epiphytes, hemiepiphytes, and vines. All figs are pollinated by fig wasps, which resemble thrips in their tiny size, feeble flight, and vigorous breeding. But whereas similar Philippine mahogany trees can coexist by using a communal set of several thrips species at separate times, similar figs can coexist because each has one wasp species of its own. The result for figs is as harmonious as it is for Philippine mahoganies. Fig species don't have to compete for pollinators and pollinator species don't have to compete for figs—albeit some wasps parasitize the system by stealing from figs without pollination!

A fig wasp's life is more unusual than a thrips'. Whereas most pollinators transfer pollen unintentionally, a female fig wasp gathers pollen from one plant into pouches on her sides and later methodically delivers it to flowers on another plant. Delivery takes total commitment. Several hundred tiny brown flowers lie crammed into the fleshy inner rind of an unripe fruit, called a syconium. To reach the flowers, a female pushes through the snug gap in the syconium—an action so strenuous that it strips off her wings and much of her antennae. Slanted protuberances block her departure. She injects eggs into some of the concealed flowers, fertilizes many of the other flowers with pollen, and dies. (Commercial figs, being sterile, do not contain dead wasps.)

Each larva feeds off the ripening seed in its flower cradle. The wingless adult males emerge first from the flowers into the inky syconium core. With their mandibles, males slice holes into flowers containing females. After mating they help each female emerge from her flower. The males then cooperate to tear an exit out of the syconium, crawl outside for the first time, then drop from the fig, dead. The females gather pollen from the remaining flowers in the syconium, leave by way of the males' exit hole, and seek fresh blossoms inside another unripe fig of the same species, just as their mothers did. The coffin of one generation becomes the womb for the next.

Pollination and Rainforest Stratification

Nimal and Savi Gunatilleke are experts on canopy pollination in Sri Lanka, an island nation off the tip of India. Both professors at the University of Peradeniya, this married couple focuses on the Sinharaja Biosphere Reserve, Sri Lanka's last major tract of lowland rainforest. Years prior to meeting them, I explored this gorgeous region on a search for spiders. I learned then that Sinharaja abounds with unique species.

Savi and Nimal have their own way of getting into the canopy. They hire intrepid village climbers to raise ladders, section by section, up trunks of the slim native trees, tying each in place. In the crown, the climbers top the ladders off with a bamboo platform. Sri Lankan tree crowns tend to be small, so the scientists can reach flowers effortlessly from this central location.

I joined Savi and Nimal in a tree, a dipterocarp, to watch the bustle of honeybees reaping nectar and pollen. Dipterocarps dominate Sri Lanka as they do Southeast Asia. But canopy blooms in Sri Lanka are not of the general, forest-wide sort documented by Appanah in Malaysia. Many Sri Lankan trees bloom predictably or on schedules independent of one another. Although at least one group of dipterocarps, pollinated by bees, does flower in a tight sequence, there is no evidence of pollination by thrips or similarly prolific insects.

Bees are dominant pollinators in Sri Lanka, as they may be for most tropical and temperate habitats. Most rainforest bees (such as orchid bees) live in solitude. Sociality occurs in bumblebees, stingless bees, sweat bees, and a few others, all represented in Sri Lanka. Asia is remarkable for wild honeybees; Sri Lanka has both dwarf and giant species.

In the famous dance language of the honeybee, a returning bee dances for her nestmates to inform them of the distance, direction, and quality of the flowers she's found. This doesn't work so well in multitiered tropical forests, because the bee cannot communicate height in trees. Instead of dancing, stingless bees do honeybees one better by depositing a scent at intervals along the route, guiding recruits to the exact flower location in the canopy.

There is a loose segregation of pollinators by rainforest stratum in the tropics. Small bees are most active in the understory, and large species, such as honeybees and bumblebees, pollinate a greater proportion of sunlit canopy trees. Birds and bats are also prevalent upper-canopy pollinators worldwide (though in the lowlands of the New World, the flowers favored by hummingbirds occur mainly at understory levels near treefall gaps, providing a cheerful peppering of color at ground level that is largely missing from the Old World). Many species of birds, bats, and large bees fly swiftly over great distances, and may prefer to stay above the forest for long sojourns.

As we saw for some of the mass-flowering trees of Asia, an advantage of fast-flying animals is the distance at which they can travel to floral advertisements. Many of these trees are emergent species, which, for an enhanced effect, bedeck themselves with densely packed flowers for a short time in a "big bang," turning crowns into giant, colored beacons. Some of the species drop their leaves before flowering, which makes their displays all the more dazzling. Even Old World bats called flying foxes, with their bug eyes and poor sonar, may spot the pale humped tops of favorite dingy-flowered trees far off at dusk, if the pungency of all the flowers doesn't draw their attention first. Bare of leaves, the trees also offer more elbowroom for bats visiting them—some of which have wingspans of one to six feet.

The problem with the "big bang" strategy isn't in luring pollinators, but in getting them to leave the bountiful tree to convey pollen to the next tree. Departure is hastened in various ways, as when gangs of courting males harass female solitary bees into departing, or when aggressive species drive away other kinds of pollinators. Individual trees of a species may also have different daily schedules of nectar production.

Some plants, particularly in mid-canopy levels, appear to bloom continuously throughout their adult lives, a few flowers at a time. Perhaps, shaded by trees above, they don't have the energy resources for costly "big bangs." The plant's offerings are predictable enough to be worth seeking, yet pollinators are compelled by the meager repast from each plant to visit several individuals of that species, and hence they cross-pollinate them.

Examples of this type of pollinator include the New World *Heliconius* butterflies, which resemble jet-black ribbons blotched in warm colors, and the Old World *Euploea,* related to the monarch.

A *Heliconius* butterfly rouses itself early to navigate standard routes to *Psiguria* lianas and other favored plants. Each individual learns the location of choice plants from long experience in the forest. Every day new flowers open. Once a butterfly checks out its plants' orange flowers for that day, it keeps on the move throughout the morning, attending each flower at intervals. Between visits, the flowers have time to replenish their nectar. This "trapline" strategy also works with some bees, birds, moths, and bats, but it requires steady flower production. When a *Heliconius*'s daily appraisals reveal a shortage of blooms, it may choose to stay at one flower, not budging even if airborne competitors batter at its wings in frenzied attempts to take the food.

The Slaughter of the Seeds

Mass flowering in dipterocarps is one extreme in plant reproduction strategies. It may merely be a quirky accident of the history of dipterocarps, but today far more than just dipterocarps participate in the flowerings. Why have so many other rainforest trees in Asia adopted the same unorthodox breeding rhythm as the dipterocarps? The explanation lies in another area of ecology: seed dispersal and the survival of offspring.

Whether pollinated by thrips or beetles, birds or bats, the flowers in the great trees of Southeast Asia soon wither, their petals fall away, and seeds swell in their place. In a matter of weeks, the seeds begin their own descent. Ranging from one to six inches in length, the shuttlecock-like seeds of dipterocarps

spiral to earth, as botanist Henry N. Ridley determined in 1930:

> When one of these fruits falls from a lofty tree, it drops vertically for a distance of 5 or 6 feet, and then the tips or as much as half of the wings spread outwards, and the fruit, still falling vertically, commences to rotate rapidly. This lessens the speed of the fall, and allows the wind to act upon it and drift it to a considerable distance from the tree before it reaches the ground.

The winged seeds work best in the biggest dipterocarp trees, where they can spiral unimpeded for some distance before hitting other foliage; many of the dipterocarps in the main canopy have seeds without wings. Still other, nondipterocarp trees drop seeds or nuts in an assortment of shapes, or pods that resemble overgrown beans. After a mass flowering, the earth turns into a spilled treasure chest of seeds, each one packed with nutrients and energy reserves for the seedling. Tree populations like this that cast off huge numbers of seeds at once (especially at unpredictable intervals) are said to "mast."

Animals have a field day, gorging themselves on the seeds before and after they fall from trees. Many Southeast Asian birds migrate for miles to areas of peak seed production. On the forest floor bearded pigs cover marathon distances in noisy herds of tens or hundreds. Meanwhile, silent swarms of weevils and bark beetles gnaw on seeds in the canopy or on those freshly fallen to earth. But whereas many insects merely blemish seeds, the fattened vertebrates destroy them. Parrots crack seeds in their beaks; green pigeons and forest pheasants grind them to pieces in their gizzards; squirrels chomp them between their molars. Digesting a seed diet, animals like these are called "seed predators."

Humans likewise can be seed predators, even of dipterocarps. As forester Frank G. Browne described in 1954, seeds of masting Philippine mahoganies and some other dipterocarps have long been important to diet and commerce in Sarawak, as they are today:

> The fat for which these seeds are valued is a tallow-like substance. . . . In Europe the product is used largely in the manufacture of chocolates, and also of candles and soap, or else the nut meal may be made into cattle-cake. Locally most native tribes eat the fat as a nourishing delicacy mixed with rice. . . . A good nut-year is a great event in Sarawak and provides a literal windfall for the native tribes.

He reports that from five thousand to sixteen thousand tons of the seeds (known in the trade as illipe nuts) were harvested in masting years.

We may think hordes of seed predators would be

ABOVE, LEFT TO RIGHT: **In contrast to the exacting relationship between a fig tree species and its pollinating wasp, the *Elaeocarpus amoenus* tree of Sri Lanka has small, nodding flowers that attract a plethora of flies, bees, butterflies, wasps, and other insects, any of which might transfer pollen from plant to plant.**

catastrophic for masting trees, but seeds in a mast face better odds of survival than if they matured out of sync with other trees. In some Asian species, a minority of individuals do bloom in "off years." When that happens, not only are few seeds produced, but they tend to be wiped out by famished predators. Successful reproduction in these species occurs mostly in mass-flowering years.

Masting resembles a desperate gambit by an outflanked army: a suicidal advance in which most die so that a few make it through. Unable to fight seed predators, trees carpet the forest with offspring so abruptly that it's impossible for seed predators to eat every one. Before enough pigs and parrots (and people) arrive to destroy everything, the seeds they've missed have sprouted and the feast is over.

Ordinarily, seeds that fall into the dense cluster under their parent are killed by predators and parasites that converge there (see Chapter 2). But if the rain of seed from many tree species is so dense that edible seeds lie everywhere, then where a seed falls may make little difference to its prospects. This may be one reason for the unusual clumped distribution of many tree species in Asia.

Masting helps explain the success of temperate oaks and tropical trees that form monodominant stands. However, in tropical rainforests that have many tree species, masting like that in Malaysia is an exception, not a rule. Outside of Southeast Asia, a majority of tree species breed independently. Although not all species set seed every year, there are usually many seeds of some sort or another ripening, year-round. This supply permits undiscriminating species of seed predators there to maintain populations so large that no one tree species could glut the food market with seeds.

In tropical regions of high tree diversity, only when many tree species reproduce together can the effect of masting be intensified enough to overwhelm seed predators—and this is known to occur only in Asia. The effect is enhanced by the precision of masting. The first dipterocarps to bloom develop seeds slowly, holding on to them while later species catch up. Though the flowering of all these species lasts months, the trees drop their seeds in synchrony over just a few weeks. Any other tree species whose seeds ripen at the same time can benefit from the food glut. As a result, the diversity of participants in Asia's mass flowering (and the masting that ensues) has probably multiplied over evolutionary time.

Not all tree species that join the mass flowering do so to avoid seed predation. Whereas most

ABOVE: Ripe seeds of *Shorea gardneri*, a dipterocarp tree. *(Photo: Darlyne Murawski)*

FACING PAGE: As large as a house cat, the Indian giant squirrel (*Ratufa indica*) is a seed predator.

dipterocarps and other mass flowering trees have defenseless seeds, a few species have toxic seeds that predators won't eat anyway. These trees may make use of the same temperature cue as the dipterocarps by coincidence.

The reproductive strategy that augments tree diversity in Asia wreaks havoc on animal diversity. Around the world, most trees that provide reliable food for tropical pollinators and fruit- and seed-eating animals position their crowns in the upper canopy. Many Asian dipterocarps—which themselves seldom provide animal food—are emergent species that tower over these normal-sized trees. Peter Ashton of Harvard has evidence that since dipterocarps are much more numerous than are the emergent trees of Africa or the New World, their shade suppresses the growth and reproduction of upper-canopy trees more than elsewhere, so relatively few Asian animals can specialize on the pollen, nectar, or seeds they produce. In comparison, tropical American forests have, for example, many more individuals and species of nectar-seeking birds and bats and of seed-eating songbirds.

Moreover, Asian vertebrates tend to be larger than their New World equivalents. Consider that hornbills are bigger than toucans (known among Latin Americans as "flying bananas" for their absurd silhouettes in the air), and flying foxes are bigger than any New World bats. Perhaps this is because, metabolically, large species can get by on inferior foods when times are lean.

Clearly, the rhythm of tree reproduction, by affecting the production of forest foods, has repercussions throughout the canopy's intricate tapestry of life.

CHAPTER NINE

TREETOP GAMES BETWEEN PLANTS AND ANIMALS

I heard one of the most remarkable sounds to reverberate among the trees of Asia while hiking in an Indonesian rainforest. My guide stopped me with a raised hand, his face taut with excitement. We listened. A low grumbling, like a drunk's wordless growl, slowly crescendoed into angry bellows. I didn't stop long to ponder what could make such a racket. I ran hard toward the roar, which persisted, unabated. It came from farther away than I had guessed. As I raced on, my vision blurred with sweat. I wiped my eyes, and there, high in the canopy, fur rippling like orange fire, sat the great ape. I was hearing the "long call" uttered but once every few days by an adult male orangutan.

The call began to peter out soon after we found the orangutan. Nestled in a bank of sunny treetop foliage, he fell silent. He watched me shrewdly for a moment, then gazed in another direction, feigning boredom. That was my interpretation. Perhaps we humans were uninteresting. From all the fracas, I had expected to find two males on the verge of battle. Only later did I learn that the unnerving cry probably serves to warn other males where a caller is, so as to avoid confrontations, not to cause them. Orangutans, it seems, lead solitary, wandering lives, seldom fighting.

My eyes on the ape, I lay back on the earth, the odor of decaying leaves rising from beneath me. Asrab, my guide, rested nearby. What a welcome to Indonesia's Gunung Palung National Park, I thought. As with most unspoiled sites, it had taken an effort to get here. From Pontianak on Borneo's west coast I had taken a motorboat along the coast and up a system of rivers (with stops to have papers checked by inquisitive police and forestry officials). After a night on a wood floor in a village, a drone of Indonesian voices (talk of fish) lulling me to sleep, I hired a dugout canoe and help to pull it (I had been told to bring two people; the villagers insisted on four). For two days we dragged the leaky dugout weighted with gear up narrow streams, our feet slipping on a mud bed made treacherous with roots. (Only one villager and my field assistant did the work; the other three usually trailed behind.)

What an incongruous sight I faced at the end of that journey. Arriving late the second day, I found a field station with library, labs, and private sleeping bungalows, each with a sandy streamside beach. Harvard biological anthropologist Mark Leighton oversaw construction of the station in 1986, sponsored by the Indonesian government.

Mark's field team studies fruit and seed consumption by rainforest birds and mammals, the most ambitious research of its kind. Some animals are seed predators, which kill seeds by eating them. (Most seed predators are the equivalent of nectar and pollen robbers, which cheat plants by taking food from flowers without pollinating them.) But animals like orangutans, called frugivores, help plants by carrying off seeds, often digesting only the swaddling fruit layer around the seeds. Though the diversity and abundance of frugivores in Southeast Asia may be comparatively low because of masting (see Chapter 8), apparently, even in Asia, enough plant species produce fruit on a regular basis to supply food for these animals year-round.

From the time I arrived at the station, I'd kept my eyes peeled for big furred or feathered frugivores in the act of eating a meal. The loud male orangutan did not let me down. After lounging in the sun and pointedly ignoring us, he descended to another tree—orange pelt turning lackluster in the understory—where he gingerly plucked tan-colored fruit and scrupulously ate each morsel. Suddenly having enough of us, he stopped feeding, shook the tree, snapped a branch, and hurled it in our direction. He retreated into the canopy, his beautiful mane flowing out of view.

Plant Sex and Animal Go-Betweens

The orangutan's appetite for fruit reminded me that plant reproduction entails more than pollination. Though plants grow so as to reach light and some climbers meander animal-like through the treetops, plants never move in search of mates. They depend on outside agents—

PAGE 152: In Borneo, a *Ficus stupenda* fig draws in a multitude of vertebrate species that savor its fruit, such as the long-tailed macaque (*Macaca fasicularis*).

PAGE 153: Capable of grinding up the fruit's tiny seeds in its crop, the thick-billed green pigeon (*Treron curvirostra*) is one of the few fig seekers in Borneo that is a predator rather than a disperser of fig seeds. *(Photos: Tim Laman)*

Crowned woodnymphs (*Thalurania colombica*) are hummingbirds whose territoriality may restrict the gene dispersal of the plants they pollinate.

water, wind, or animals—for sex and for the dispersal of their offspring.

Higher plants disperse in two stages, first as pollen grains (which contain the sperm), then as seeds (which contain the embryos). Once the pollen is dispersed, it can fertilize the immobile ovule, which is held by the mother. The mother nurtures the embryo, fortifying it with the food and armored coats it will need to survive in the hostile outer world.

Two-stage dispersal has its parallels among animals. In most animals, the males or the females or both sexes move about to find suitable mates, thereby dispersing their eggs and sperm. The offspring that ensue may be nourished and protected by their parents or they may be left to fend for themselves. Regardless, at some point they typically disperse to find suitable unoccupied habitats, much as the seeds of plants disperse.

"I think of dispersal in terms of genes," Darlyne Murawski (who investigates plant population genetics at the University of Massachusetts at Boston) declared on a drive in Panama's Soberania National Park. Why genes? Biologically, the success of an individual tree (or an animal, for that matter) depends on how well it casts its genes over the landscape to form part of the next generation. The dispersal of genes presents a critical problem in the tropics, where individuals of a given species typically are few and far between.

A tree, sowing pollen far enough to reach others of its kind, fertilizes some of the flowers of these other trees. If the tree is bisexual (both male and female), it may in turn have its own flowers fertilized by incoming pollen from them. All the resulting seeds are related equally to that tree genetically, regardless of whether or not they actually grow on it. Thus, even if seeds simply drop to earth, a tree has moved its genes as far as its pollen was carried to other trees.

What are the consequences of adding seed dispersal to pollen dispersal? In tropical forests that are diverse in tree species, seed dispersal allows trees to reach the unoccupied terrain between adult trees as possible sites for some of their offspring. "But, perhaps more importantly, by shuffling a species' genes around one more time, seed dispersal increases genetic diversity in an area. Such diversity is vital," Darlyne said. "At its root, biodiversity simply is genetic diversity. When many scientists speak of biodiversity, however, they mean only species diversity. While it is obvious that species must differ genetically, populations within a species vary genetically, as well."

Methods now exist to determine the parentage of seeds. These methods allow Darlyne and her colleagues to describe the mating behavior of trees, and to establish the distances between parent trees (pollen-dispersal distances) and the distances between established seedlings and their mothers (seed-dispersal distances). Most of what we know today concerns the trees of Central America, though Darlyne has begun pioneering work in the Asian tropics.

Darlyne stopped the Jeep and headed into the forest, slingshot in one hand. On a hillside we came upon a tree whose trunk, pale and barrel shaped, had been visible miles away: the cuipo, a relative of African baobabs. Darlyne walked up to the tree and pressed her ear to the bark of the two-yard-wide trunk. She motioned for me to do the same. We thumped the tree with our palms. The soft, water-retaining wood within echoed back.

Then she pointed to cuipo seedlings around the parent tree. Among the healthy green shoots were bleached sickly ones, doomed to an early death. "Sometimes, when pollen dispersal fails, we find albinos like these," she said. If no pollen reaches a cuipo flower from other trees, the flower can still set seed by the last resort of self-pollination (called "selfing"). Albinos and other genetic anomalies commonly arise from this type of inbreeding, but not all the seedlings are defective, so the tree may still successfully reproduce.

Cuipo adults occur either as scattered individuals or in clumps. Darlyne finds that clumped trees have the fewest selfed progeny. Most receive pollen from other individuals (they "outcross"). But isolated cuipos like the one before us receive little pollen from other trees. "Not all trees flower each year," she added. "If a tree in a clump blooms when none of its neighbors do, getting pollen from other trees will be more difficult than it looks. This means the tree may pollinate itself heavily that year."

Picking up a stone, Darlyne took aim at the tree crown with her slingshot. On the third try, down tumbled a branch. She stuck it in a bag. Though she climbs trees, she finds it's not the fastest way to collect samples. "I'll check the genotype of this tree back in the States by allozyme analysis," she told me. By comparing its genes with those of its seedlings and of its adult cuipo neighbors, Darlyne can determine whether a tree has been selfing.

Because many tropical trees are rare, some scientists once argued that selfing would prove to be the norm. The reverse turns out to be true. Most tropical trees seldom if ever self-pollinate: either their own pollen is rejected by their own flowers, or individual trees are single-sex, some being male and others female. Moreover, Darlyne, during work with Jim Hamrick of the University of Georgia at Athens, has shown that while rare tropical trees have significantly less genetic variation than common ones, they're still more diverse genetically than had been anticipated, indicating that their genes must be moving long distances. There is now genetic evidence from a couple of tree species of pollen dispersing farther than a quarter-mile.

"To preserve biodiversity, it's important to conserve diversity within a species," Darlyne said. A Noah's Ark conservation strategy won't work: small, isolated populations are often characterized by low genetic diversity, frequent inbreeding, ill health, and an increased likelihood of extinction. "Most tropical trees are rare, but genes are moving long distances, so populations cover big areas. The number of interbreeding individuals probably still has to be large to keep the species going," she told me. "Their survival demands large tracts of rainforest."

The Reliability of Animals as Dispersers

To move genes long distances through the tropical canopy labyrinth, the choice of dispersal agent—usually some sort of animal pollinator or frugivore (seed disperser)—is crucial. Clearly, plants will go to extremes to make sure animals depart with fruit or pollen. Writer Alex Shoumatoff recounted one extreme example of deceit pollination in the Congo:

> White, trumpet-shaped blossoms . . . dripped from a twining liana in the birthwort family. Flies were induced to enter the trumpets by a carrion smell; were trapped inside, by retrorse hairs; and were released within twenty-four hours, smothered in pollen, when the hairs wilted.

But how effective are such animals as dispersers of plant genes?

From a plant's point of view, the ideal pollinator would attend to the flowers of its species and no other—the perfect servant, custom-made for the species' needs. But plants such as the birthwort relative attract a range of species. In the tropics, only fig wasps are faithful to single plant species (see Chapter 8). In fact, a widespread rainforest canopy strategy is to display small generalized flowers that draw in a kaleidoscope of wasps, bees, butterflies, flies, and other insects that visit many plant species—though some visitors may be better than others at successfully moving the pollen between plants. Many of the plants with this generalized strategy are trees of the upper canopy that

When a Hanuman langur (*Semnopithecus entellus*) chews on the wind-dispersed fruits of the *Terminalia* tree, it most likely kills the seeds. The langur diet consists primarily of leaves, including those of plants extremely toxic to most animals (India).

have separate male and female individuals to assure outcrossing.

In complex rainforest environments, some degree of specialization usually is necessary because, given the diversity of flora, chance alone would not guarantee that floral visitors would transfer pollen to the same species. As a result, flowers typically conform to one of a few broad patterns, or syndromes, fashioned to attract and reward a certain class of pollinators. Vertebrate syndromes (usually involving birds and bats) are more common in the tropics than in the temperate zones. Yet the total number of insect pollinators must be astronomical by comparison, given that nine out of ten equatorial plants show an insect pollination syndrome. For example, small tubular flowers that are pale, fragranced, and nocturnal will probably entice moths, whereas those that are orange, weakly scented, and diurnal more likely attract butterflies.

Each of these syndromes contains its own special cases. Among the most distinctive moths, for example, are the hawkmoths. The most spectacular of their flowers may be the angraecoid orchids of Madagascar, whose narrow nectar spur tubes extend up to fourteen inches! The hawkmoths hover at flowers as do hummingbirds and use proboscises resembling needle-thin soda straws to reach the nectar. Each angraecoid orchid species is pollinated by a hawkmoth species whose proboscis length matches the nectar-tube length. Between meals, the hawkmoths roll up their proboscises like a fire hose.

In all these syndromes, several pollinator species are likely to share several plant species. Plants can still neatly avoid having their pollens mixed up. For example, different angraecoid orchids produce pollen that, because of its location on the flower, sticks (in a packet) to distinct parts of a hawkmoth's head or thorax, and so ends up deposited only on flowers of the right species.

Reliability at the second dispersal stage (the seed) is trickier to evaluate. After all, from a plant's standpoint, a pollinator's job is simple: carry pollen

The vermilion capsules of *Sterculia* fruits open to reveal flesh-coated seeds that birds adore (New Guinea).

from one flower to another of the same species. The best target for a seed—a likely site to germinate and mature—isn't easy to discern in a complex rainforest. Indeed, it may be next to impossible to predict for long-lived plants how the forest will change as the plant grows.

In the rainforest, wind dispersal of seeds is much more common than wind dispersal of pollen—presumably because a seed is more likely to contact a good germination site by chance than a pollen grain is to contact the stigma of a flower of the right species. But seeds do not merely fall: they gyrate, undulate, tumble, or (as in dipterocarps) helicopter down, or they waft along on milkweed-like fluff. Paper airplanes are crude by comparison. "Consider the flying abilities of a seed from one of Panama's *Pithecoctenium crucigerum* lianas," Tom Croat, of the Missouri Botanical Garden, once said to me. "It's a superb glider, often wandering aimlessly through the forest understory and with such sensitivity to obstructions that it swoops under or over them."

Orchids—mostly epiphytes in the sunny canopy—show a different wind dispersal strategy, releasing as many as two million microscopic seeds from one flower. Joining the rainforest's aerial plankton of thrips, fig wasps, and others, these particles travel farther than winged or fluffy seeds and end up coating everything. Although an infinitesimal fraction of them do reach fertile sites by chance, the strategy (and, to some extent, other wind-dispersal strategies) does not allow a seed to bear weighty food stocks for the developing seedling. Orchid seeds must bank on lodging next to the right fungus, which will dispense to it most of the nitrogen it needs for growth.

Some tropical seeds are thrown explosively from their capsules. With a sound like a rifle discharging, the New World sandbox tree zings its flattened seeds so fast that they have been known to injure people.

Most temperate flowering plants have light, wind-dispersed seeds, or clinging, burr-like seeds. Yet nine out of ten tropical rainforest species have bulky, well-stocked seeds with an edible coat, or "fruit." Some frugivores swallow seeds with the fruit and defecate them unharmed; others spit them out and swallow the pulp alone. Any fruit or feces discarded with the seeds also can serve as compost for the seedling.

Pollen has the advantage of weighing too little to be a burden even to the smallest insects. Seeds—stocked with much of what the seedling will need—are ballast worth discarding at the earliest convenience. No wonder, then, that vertebrates are overwhelmingly the animals of choice in tropical fruit dispersal. Bribed with fruit, vertebrates haul seeds farther than insects ever could.

Many mammals take days or weeks to digest fruit, so seeds may be deposited far from the parent plant even when excreted by slow-moving creatures. Thus flight is often superfluous for seed dispersers, which include primates, rodents, and other canopy wanderers, as well as birds and bats. Pollen, in contrast, must reach flowers on widely scattered plants of the same species before the pollinator grooms it off. Perhaps this explains why pollinators almost always fly. Exceptions occur wherever appropriate animals don't exist. For example, where flower-visiting bats are rare, less-effective flightless mammals have been implicated as pollinators. Lemurs—primates closely related to monkeys, apes, and humans—pollinate some rainforest trees in Madagascar.

In the Amazon, where some forests are inundated annually by swollen rivers, many trees cast fruits into water where they raft downstream. Fish often aid their dispersal. Some fish (even a few piranhas) specialize on fruit, swimming in the understory of the flooded forests to seek palm fruit and others by smell. The fish disperse some kinds of seeds, but they kill others: when the drying pods of rubber trees burst open with a firecracker-like bang, the fish gather where the seeds splash into the water and grind them to pieces. One impressive frugivore, the tambaqui fish, can weigh seventy pounds by the time the floods recede. Thereafter the tambaqui subsist mostly on fruit-culled fat reserves until the high-water season returns.

Dispersal distance varies. Conditions near pioneer trees become intolerable to offspring as their parents fill in the canopy gap. To compensate, pioneers have huge seed shadows; their tiny seeds are widely disseminated and remain dormant in the soil until gaps open overhead. Then they germinate in response to light or temperature. Most trees in "mature" forests—climax trees—produce fewer seeds that sprout days after dispersal. The developing seed incorporates food reserves so that the seedling can survive in dim light. Pioneer seeds, which sprout in an environment steeped in sunshine, need less food and so are often tiny and wind-dispersed.

A climax tree's seeds seldom travel as far as those of a pioneer. Still, some animals do the worst possible thing after eating fruit: they drop seeds without taking them anywhere. For tropical trees, at least, seeds and seedlings near their mother tend to be killed by herbivores or diseases. The result appears to be a catch-22 for the plants: they must first entice frugivores, then drive them away—with seeds.

In some cases, the architecture or leafing pattern of a tree may compel visitors to carry off their fruit meals (although it's yet to be proved that such tree features evolved "for" that purpose). Trees with sharply angled or weak limbs may not provide comfortable spots where weighty animals can relax and eat. Fear may also encourage prompt departure. In trees that come to fruit after shedding their leaves, eagles can readily discern monkeys and other frugivores, and boas seeking prey may lurk in any tree crowded with visitors. Wary animals respond to potential danger by grabbing fruits and retreating to less risky places to eat.

At exposed sites, winged frugivores that can flee quickly from enemies may be less at risk than flightless species. In fact, some flying foxes and parrots select for their communal roosts emergent trees that tower above the canopy. Moreover, they strip leaves from branches, giving a clear view of impending danger.

A male bushy-crested hornbill *Anorrhinus galeritus* gives fruit to a female sealed in a tree hole while rearing her brood; only the slit-like opening into the nest is visible. Sealing the nest with mud and excrement protects the young from predators and presumably creates a more uniform environment within (Indonesia).

Seeds that have been spat out or defecated, or have fallen untouched below a parent plant, can be moved farther by secondary disperser. For example, ground-dwellers like Asia's red jungle fowl, a relative of the domestic chicken, are seed predators that can also disperse seeds. Squirrels and the tropical American rodents called agoutis bury seeds to eat later. As happens with temperate squirrels, if one of the animals forgets where it cached its food, the interred seeds may germinate.

The body size of secondary dispersers need not be as limited as that of the frugivores that have to climb trees; thus, the fruits dispersed by ground-dwellers can occasionally be as large as a bowling ball. In some African rainforests, certain fruits may need to pass through an elephant so that the seeds can sprout in the fertile dung. Germination of the seeds of some tropical American tree species that are now in decline may once have been expedited through digestion by behemoths such as giant sloths and elephant-like gomphotheres, which became extinct centuries ago.

The ideal frugivore would do more than remove seeds from a plant's vicinity. It would deposit the seeds where they have a chance not only to germinate but to grow (for years or centuries to come) to maturity. For every plant species, each site presumably offers a different chance for success, depending on moisture, light, and so forth. For some plants, just about any place will do; but for others, like epiphytes needing the right soil and branch slope, good sites may be hard to find.

A bee gets a nectar reward not only for taking pollen from one flower, but for delivering it to the next. In contrast, though a plant rewards frugivores with a fruit meal for removing seeds, it can offer no bonus to animals that deliver seeds to good sites. This means that lots of seeds go to waste. Such is obviously the case, for example, with tropical American oilbirds, which resemble nightjars. They harvest fruit at night, echolocating like bats, but with audible clicks. Oilbirds nest in colonies in caves, where they defecate most seeds to certain oblivion. But enough of the seeds apparently land en route to the caves to propagate oil palms and laurels, the bird's preferred food trees. (Plump juveniles, reared on fruit until heftier than their parents, are so greasy that Venezuelans once boiled them to extract lamp oil.)

Perhaps the best a plant can expect from its vertebrate frugivore is that it leaves the seeds in the correct general habitat. Many mistletoes have soft fruit relished by many birds, but some have tough fruit that can be eaten only by flowerpeckers in the Old World and euphonias (a type of tanager) in the New.

In the Nilgiri Hills of southern India, Priya Davidar, a professor at the University of Pondicherry, showed me why these mistletoes can rely on their birds for seed dispersal. We had stopped to watch flowerpeckers flit from mistletoe to mistletoe in a tree when she asked me to take a look with her binoculars. "See the strands of seeds hanging from those branches?" She told me these had been defecated by flowerpeckers. As with many fruit-loving birds, euphonias and flowerpeckers have short digestive tracts that strip seeds of pulp quickly—in thirty minutes. This speed permits the birds to eat enough nutrient-poor fruits each day to achieve a balanced diet, and it leaves the seeds unscathed. The birds void the seeds in sticky strands, usually wiping them onto tree branches perfect for germination.

Incidentally, Priya has documented that some Indian mistletoes have flowers resembling their fruits. When a flowerpecker mistakenly pecks at a flower, the flower bursts, spilling pollen on the bird's face. Thus the same bird species acts both as the mistletoe's pollinator and as its seed disperser.

It is easy to see that mistletoe birds serve their plants well. This is true of at least a few other frugivores, like ants, the only invertebrate seed dispersers; certain ants discard epiphyte seeds in their treetop nests, which sprout to form "ant gardens" (see Chapter 6). But what of the majority of frugivores in a tropical forest?

The Canopy Cornucopia

After the male orangutan that Asrab and I watched in Gunung Palung National Park moved on, I examined the fruit he

Mark Leighton apportions Bornean fruit into dispersal syndromes. Only hornbills eat the large bisected fruit at top center (they can gnaw past the thick white rind) and the orange fruit in the top left corner; monkeys consume the pale orange or green fruits in the upper right; rodents disperse the dull nuts at the bottom right corner; the fig spray at bottom center is widely popular; and small birds favor the brightly-colored fruits at the left.

had been eating, which he shook down from the tree in his departing tantrum. They were thin-skinned light-brown globes a bit larger than grapes. At the station I handed one to Niki Yonkow, one of the student managers of the Indonesian team's field projects. "An orangutan ate these?" Niki asked, studying it from all sides. She bit into it, and gestured that I do likewise. The pulp was orange, delightfully sweet and aromatic. "You see, if a primate here likes a fruit, chances are you will, too. We share similar tastes."

Thanks to these primates, the rainforests of Asia present a more bountiful harvest of fruits to delight the human palate than any other location on earth. Village markets overflow with mangosteens, rambutans, breadfruit, jackfruit, lychees, mangoes, gandarias, langsats, star apples, salaks, and other local delicacies. Of one spiny Malayan species that ripens to "about the size of a child's head," Swedish scientist Eric Mjöberg declared in 1930 that

> one thing is certain, our first meeting with the durian fruit is a memory which we carry with us to our life's end . . . its smell is something between perspiring feet and rotten eggs. . . . The natives are passionately fond of durian, and race to the great forest tree when the fruits are ripe and ready to fall.

Singapore has laws against taking durians to hotels and other public places because of their gag-inducing odor, yet street vendors can be swamped by customers methodically sniffing, scratching, and thudding durians in age-old tests to select the finest among them. Western visitors can be dubbed "honorary Asians" if they claim to relish this strange fruit. Rainforest mammals seem to feel the same way. Even tigers will wolf down durian meat.

But Mark Leighton of Harvard believes most Bornean fruits match the palates of specific groups of dispersers. To him, fruits sort into syndromes as neatly as flowers. He defines rodent fruit, civet fruit, bat fruit, hornbill fruit, primate fruit, bear fruit, and more.

As it turned out, the orangutan I had seen had been eating *Microcos* tree fruit, one of the many favored foods of resident primates. Primate fruits, Mark finds, tend to be yellow, orange, or brown, with a rind or husk enclosing a juicy pulp and one or two seeds. *Microcos* fruits fit the syndrome perfectly.

Fruits taken by small birds in the tropics tend to be small and often black, red, or a mix of both colors. Usually they are sweet, but they offer little nutrition, since many such birds get the protein they need from insect prey. Small birds have a harder time removing husks than do agile-fingered monkeys; the fruits they eat either lack a husk or have husks that open by themselves. As in temperate zones, most of these berries are eaten by many bird species.

Other, larger fruits tend to be sought by more restricted groups of birds, like the hornbills or birds of paradise. These fruit specialists—which are unique to the tropics—have high expectations of their fruit meals: they are gourmets rather than gourmands. To exclude the wrong clientele (presumably less-reliable dispersers) from these costly, nutritious feasts, the fruit are often encased in husks that smaller birds cannot remove.

But if large-bird fruits are so desirable, what stops a monkey or an ape from opening them for a snack? Later in my stay at Gunung Palung I hiked with Scot Zens, a Harvard undergraduate pondering this question. Scot had been at Gunung Palung a year. "On the surface the hornbill fruits around here look perfect for primates, but they avoid them," Scot said, pausing to pull a leech off his toes. (Scot is a rare breed of Westerner who likes to go barefoot in a rainforest. Fortunately for him, Asia's infamous land leeches appear to be absent from treetops.) Scot found that the fruit accumulate chemicals, called terpenes, that confer on them the bitter flavor of pine needles. This is a sensory experience that birds seem to enjoy but that mammals find loathsome.

A fruit, like a flower, may proffer a hundred or more trace compounds that endow it with a unique quality of aroma and flavor that delights some animals but makes others gag. Some of the compounds may be the same as those that plants employ to dissuade herbivores from chomping on leaves. High toxin concentrations in unripe fruit render them completely inedible to most animals, keeping even the dispersers themselves from feeding before the seeds have matured.

Positioning of fruit in the sea of forest strata can help or hinder different animals as well. Epiphytes and understory trees yield most of the fruits that sustain small birds. Emergent trees and upper-canopy vines, washed by breezes that never reach sheltered plants below, commonly have wind-dispersed seeds. Most of the heavy animal-dispersed fruits likewise come from tall trees. These fruits often are selected by the heftiest dispersers: birds and mammals needing the robust supports that lofty trees provide.

The stratification of a rainforest results in flyways between vegetational realms. Some plants hang fruiting or flowering branches into these paths so flying animals can find them. The plants separate photosynthesis and reproductive functions: certain *Passiflora* vines leaf at the treetops, but are pollinated by hermit hummingbirds that forage close to earth. As if fishing for pollinators, the plants lower reproductive shoots into the understory. Some durian and fig trees extend their reproductive branches down their trunks and along the ground. The figs may even bear fruit below the humus. These species avoid arboreal fruit dispersers altogether, in favor, it seems, of pigs or other routing animals.

A commonplace rainforest strategy virtually absent from temperate canopies is cauliflory: a tree flowers and fruits down its trunk. Fruits there are convenient for climbing mammals like civets that find the fringes of tree crowns precarious. The fruits or flowers can also be visited by bats, which veer easily through narrow flyways but have difficulty maneuvering within the luxuriant strata to harvest the food there. Yet, once they arrive at a fruiting tree, it's hardly a surprise that some tropical birds and bats are adept at walking along branches to reach less-accessible treats.

Scientific Opinions Differ

Some ecologists are not as comfortable classifying fruits into different syndromes as they are flowers. For one thing, they argue, many tropical plants have existed longer than the animals that disperse their fruits today: how could the fruits have evolved to lure the animals? Perhaps it is the animals that have evolved to seek out fruits with certain characteristics (color, size, smell, taste), rather than the fruits evolving to match their dispersers, as might also be true for some pollinators and their flowers.

Other arguments suggest fruits are often less intricately evolved to suit frugivores than most flowers are to suit pollinators. Obviously, flowering and fruiting are connected; one leads to the other. Consequently, traits of one stage may compromise the effectiveness of the other. No plant can make use of a pollinator that lives on the ground and then exploit a frugivore that patrols the topmost canopy. Nor is a plant with big flowers likely to produce tiny fruit suitable for a small bird. When ideal solutions to the two dispersal stages conflict, it's likely that the needs of the flower will win out in evolutionary time. After all, pollination is an all-or-nothing process, while fruit dispersal has at least some chance of success, no matter how inept the disperser.

Keystone Species

A paradox of the seemingly hospitable tropical environment, with its smorgasbord of fruits and other morsels, is that food is not always plentiful. For many animals feast alternates with famine. Instead of a uniform climate supporting stability, the tropical rainforest's subtle vagaries in weather can on occasion produce wild fluctuations in flowering and fruiting. The trivial cues that many tropical plants rely on to synchronize reproduction simply are not reliable. Masting—simultaneous production of seeds—in many Asian trees is just one drastic example (see Chapter 8). On Barro Colorado Island, the biological reserve in Panama, many canopy plants rely on a relatively dry spell from January to March as their signal to gear up for flowering. But in 1971, rains continued unabated, causing flower and fruit production to plummet. Many ground and canopy animals went hungry, as botanist Robin Foster documented:

> The spider monkeys . . . launched an all-out assault on food resources inside the buildings, learning for the first time to open doors and make quick forays to the dining room table, where they sought bread and bananas, ignoring the meat, potatoes, and canned fruit cocktail, and brushing aside the startled biologists at their dinner. . . . One could find at least one dead

FACING PAGE: A Singaporean *Ficus varigata* fig presents its fruit not in the congested canopy, but on its trunk, where more hungry animals can easily see and reach the tasty offerings.

The Cannonball tree from Guiana (*Couroupita guianensis*) blooms along the trunk. Large bees wedge themselves beneath the flower's hood to procure the pollen within.

> animal every 300 m along trails well away from the laboratory clearing. The most abundant carcasses were those of coatis, agoutis, peccaries, howler monkeys, opossums, armadillos and porcupines. . . . At times it was difficult to avoid the stench. . . . Fruit-eating parrots, parakeets, and toucans were not seen at all on the island during this time, and large numbers were seen migrating east toward Colombia.

Understory residents may enjoy a somewhat more constant food supply than upper-canopy species. Studies have shown that flowering and fruiting by plants in those strata are more prolonged and regular than in the upper canopy.

A problem scientists face in understanding tropical dispersal is fidelity: most fruits can be devoured by any of a wide variety of species. Animals that ordinarily find a fruit unappetizing may treat it as acceptable, if not mouth-watering, when other options are scarce. Are animals that consistently gorge on that fruit and otherwise conform to the fruit's "syndrome" really the most successful dispersers for the plant? Usually we haven't a clue.

Figs are an extreme case. For all a fig's precise alliances with pollen dispersers—fig wasps—its fruit are easy pickings for many birds and mammals. Yet when alternatives are plentiful, most animals eat fruits with less indigestible fiber and more nutrients.

Because figs can nonetheless be eaten by so many birds and mammals, they (like some palms and other plants) are "keystone species" in tropical America and tropical Asia. Just as a keystone supports an archway, keystone species such as figs come into fruit year-round, and so support the animals during times of food scarcity—when all their preferred foods are absent. At such times a fruiting fig can be the best tree in the forest around which to find vertebrates.

I visited Timothy Laman at Borneo's Gunung Palung in 1990 while he was doing field research on the dispersal and survival of hemiepiphytic figs for his Ph.D. While some figs sprout on the ground like ordinary trees, Tim's hemiepiphytic species start life as canopy epiphytes, grow roots to earth, and eventually become as impressive as the trees themselves. On a gorgeous sunny day, Tim and I climbed 150 feet to the lower part of the crown of a colossal dipterocarp tree. There he showed me a fig whose trunk grew like a distended blood vessel along the tree trunk. Around us in the canopy, its branch sprays interdigitated with the tree's.

Tim checked the fruit on the fig. They were unripe: fig wasps, I imagined, still nestled inside them. After the wasps departed, the fruits would darken, and the feeding frenzy would begin. "In a couple of weeks this tree will be a hopping place," Tim declared. "There'll be gibbons and orangutans, long-tailed macaques, flying foxes, an endless stream of small birds and squirrels, and hornbills of six species. Maybe even the bear cat—binturong—a very unusual mammal." Which of these are proficient seed dispersers and which are freeloaders? Tim smiled. "I can't say for sure . . . yet." What is sure, he added, is that in no time the canopy would be splattered with clumps of tiny seeds.

Tim agreed that figs are a keystone resource for vertebrates. "But considering that, the adult plants are not so common," he said. "They produce tons of fruit. Their seeds must be landing everywhere. So what limits colonization of new sites?" To find this out for the hemiepiphytes, Tim had to determine the best sites both for germination and for later survival. He found that seedlings of this kind of fig need canopy light. "If they have it, they grow a nice thick stalk and start putting out leaves," he said. To survive long, though, they must be in a knothole or tree crotch with lots of moist soil.

At Harvard, I had judged the six-foot-four-inch Laman to be ungainly, but he moved fluidly here, in the canopy, like a person come home. Relaxed in the warmth of the setting sun, he showed me where the seed progenitor of our fig had been excreted by a bird or mammal long ago. "It's at the point where the roots head down and the branches head up, in that tree crotch next to you."

Spider monkeys eating flowers of *Tabebuia rosea*, a tree pollinated by large bees in Costa Rica. Though it may occasionally pollinate by accident, a feasting monkey usually does more harm than good to flowering plants. *(Photo: Darlyne Murawski)*

To judge the performance of fig dispersers, Tim has to figure out the likelihood of seeds reaching good sites. "Ideal spots are small—often a square inch or two—and there are not many around. Even given millions of animal droppings with fig seeds, very few hit these targets. But once a seedling becomes established, the chance it will reach adulthood appears to be high, much higher than for a similar-sized seedling of a tree on the ground."

Our tree stood head and shoulders over its neighbors, giving us a glorious view of the rolling tropical canopy, dappled by sunset colors. This was a Borneo few had seen. Tim pointed to the east. Beyond the next ridge stretched forest in which only tribal people had ever been. From our vantage point it looked tantalizingly close. Why had Tim switched from work on the forest floor to dangling in its canopy? He said he was first drawn to climb because he missed the long view, missed seeing sunsets on the horizon. Once in the canopy, he was hooked by the challenges of research in this little-known paradise. On that note the sun's red flame began to slide beneath the earth's green disk. We began our descent amid the whispered wing-beats of bats starting their nightly prowl for insects and fruit.

FACING PAGE: In a trial experiment, Harvard scientist Tim Laman determines survivorship of fig seedlings grown in planters containing soil and other substrates at different heights. A screw palm (*Pandanus* species), an enormous epiphyte, adheres to a fork in a tree across from him (Indonesia).

In Old World tropical areas like New Guinea, throngs of certain flying foxes venture twenty miles or more each night to eat fruit, nectar, and pollen, returning daily to a tree or cave to roost in colonies. Other fruit bats engender a different pattern of plant gene dispersal, because each individual confines itself to a small territory. Bats are the most diverse group of tropical mammals.

CHAPTER TEN

A SCIENCE NEARS MATURITY

Soaring like a hawk, I skim the African treetops, inhaling floral perfumes in air enriched by photosynthesis. Before me parrots and monkeys dart from one green swell to the next, much as flying fish do when approached at sea. I remember having had dreams like this. But now, after months of bruising my limbs and suppressing vertigo just to attain a fraction of the panorama before me, Cameroon's tropical rainforest canopy freely floods my senses. I stand at one corner of a triangular "sled" suspended from a dirigible. The sled drifts lower, and in a flash I understand that all my struggles to reach the canopy may be drawing to a close. Leaning forward from the safety line that secures me to the sled, I prepare to embrace the trees.

As the canopy sled drew toward the craggy head of the lone *Gilbertiodendron* tree I had asked to visit, I forgot all the research colleagues left behind at our Cameroon base camp; I forgot even my fellow passengers, Gilles Ebersolt, the sled's designer, and Austrian physiologist Marian Kazda. I raised my net and readied myself to collect a "standard" insect sample, defined for the study under way as "the take from ten sweeps of the net through one tree's foliage." Suddenly the organic galleries of *Gilbertiodendron* enfolded me; foliage whispered past my ears and clawed my sides. I propelled my insect net forward with the full intensity of my enthusiasm.

Too late I noticed an ashen wasp nest among the tree's grass-green seedpods. I was unable to deflect my net's collision course. The rim slammed into the zebra-striped guards stationed over the nest's papery surface. Panic washed over me. My ears caught the drone of angry wasps. Before they could throw themselves outward, the sled whirled about, placing me beyond their reach.

Unaware of my near miss, Marian looked at me curiously from his end of the sled. I realized I was laughing. Some tropical wasps will attack if anything so much as approaches their nest. From my first arboreal days, blundering into a wasp colony had been one of my apprehensions. Finally it happened, and I pulled through unscathed. How differently things might have gone if I had come upon that nest from a climbing rope!

The dirigible whisked the sled through the upper canopy so swiftly I could reflect only for an instant on my luck. As I swept my net back and forth across another flank of the *Gilbertiodendron* crown, only fragments came into focus: one instant a dangling liana, the next a leaping gray lizard, then—out of the corner of my eye, what is it?—a compact, jade-green orchid.

Before I ventured to Cameroon, West Africa, in November 1991 to study treetop insects and herbivory as part of the French-run *Opération Canopée* program, I had never heard of a "canopy sled." The sled is *Opération Canopée*'s newest contribution to canopy technology. Starting with its expedition to

PAGE 170: The world's largest hot-air dirigible lowers the sled into tree crowns in Cameroon.

PAGE 171: A praying mantis oversees its canopy domain in Panama.

Meg Lowman strides across the raft in Cameroon to collect insect samples with her net as Bruce Rinker inspects the previous catch.

French Guiana in 1989, *Opération Canopée* became the first internationally publicized treetop enterprise. Only a smattering of films and writings had appeared about canopy exploration before that time.

Opération Canopée gives participants an opportunity to work alongside other scientists of many nationalities and disciplines. As had the one in French Guiana, the three-month Cameroon expedition centers around the canopy raft: mesh stretched trampoline-style within a wheel-shaped frame about forty-five feet in diameter. The frame is made of the same kind of rubber pontoons used in the much smaller canopy sled. Unlike the mobile sled, the raft is deposited by the dirigible on the canopy surface, where it remains for several days at each site. Scientists climb onto the raft by rope, and up to six people at a time can amble over the treetops at their leisure, to contemplate the canopy ocean firsthand from its very top instead of from below.

En route to Cameroon, I found out that my traveling companions—who would be there for half of my two-week stint—were prepared for famine and primitive conditions at the remote site. Arriving in the forest with stocks of American-style snack food and enough medical supplies to fill a duffle bag, they were met with a full-blown research station built solely for the expedition and fully equipped with scientific necessities: from there came the tap of computer keyboards, the tinkle of glassware, and the hum of an air conditioner. At meals the chefs served shrimp-stuffed avocados and pâtés to the forty or so scientists. Each evening during drinks and congenial discussions in a central gazebo, flanking spotlights created an aesthetic nightscape of the canopy at the forest fringe.

Francis Hallé, head of *Opération Canopée* and professor of tropical botany at the University of Montpellier in France, until recently had been best known for his scholarship on tree architecture. But since brainstorming with Gilles Ebersolt and dirigible pilot Dany Cleyet-Marrel to create the raft, his team has appeared in numerous documentaries and news stories. It could not be easy to balance the needs of scientists and journalists while managing a logistically and technically complex operation in a far-flung location, but Francis, displaying the charisma of a Jacques Cousteau, handles any complication with quiet competence and an impish smile.

During one meal, Francis diverted conversation from the raft; ideas concerned him. We talked of a subject close to his heart: reiterations in the crowns of big trees of what Francis and his colleagues call architectural models. "Chop one of these units off and place it upright on the ground," Francis said. "As I pointed out twenty-five years ago, it will look exactly like a young tree of the same species. But there's more to it than that."

Francis believes that the units are basically treelets that bud off from the parent rather than start as seeds. For certain trees, he has shown that each treelet seems to grow roots from its base high in the crown of the "parent" tree. Sandwiched between the bark and the wood, the roots cloak the tree's trunk, rejuvenating the wood with a young, strong layer and providing each treelet with direct connections to earth. The canopy roots described by Nalini Nadkarni (Chapter 4) might be a part of this system, where these roots emerge from the bark to tap into soils in the crown. In Hallé's opinion, large trees are more like colonial animals than anyone has previously imagined.

Hallé's provocative ideas have changed the way I look at a big tree. Instead of an individual, I see separate treelets competing among themselves for light and space. The crown as a whole grows dynamically, some parts prospering and others dying back, depending on the fates of the different treelets. Species lacking reiterations may grow quickly, but, in resembling the immature forms of more complex trees, they are less adaptable. "It's not surprising that such trees dominate early in forest succession, but are replaced by 'colonial' species later on," Francis says.

Some strangler figs take colonial life a step further. Rather than bud into colonies of genetically identical units, unrelated plants fuse together as they attack a host tree. "Figs are unique in many ways, and they are evolving fast," he told me. "Think about the history of land plants. First the ferns and their allies. Then the conifers and their relatives. Then the flowering plants. What will be the next great stage of plant evolution? Perhaps the group of flowering plants we call figs?"

Each morning the pilot and crew of *Opération Canopée* rose before dawn to unfold the 150-foot-long dirigible and inflate it with hot air either to transfer the raft to a new site or for that morning's sled trip.

After helping the crew with the dirigible on my ninth morning in Cameroon, I hiked into the rainforest with Margaret (Meg) Lowman, since 1992 director of research at Marie Selby Botanical Gardens in Sarasota, Florida, and Bruce Rinker, instructor at the Millbrook School in Millbrook, New York. To get an idea of the forest's structure, we made a "profile diagram" by sketching all the trees along a five-meter-wide transect to their correct heights and locations. The forest broke the "rules" for Africa: the canopy was tall (120 feet or higher) and tightly packed.

I was collaborating with Meg and Bruce on a study of insect herbivores and the leaf damage they inflicted at ground and canopy levels. Each day in Cameroon we took insect and leaf samples from the understory and, when space on the busy raft or sled permitted, from the canopy. Because most herbivores prefer soft tissue, we gauged the desirability of trees like *Gilbertiodendron* with a hole-poking device that measures leaf toughness. Then we calculated the area of each leaf using a computer scanner and estimated what portion of it had been consumed by herbivores.

Meg has gone through this procedure many times to assess herbivore damage in Australia. Once she processed all the leaves on two whole trees—220,000 in all—"just to see if it could be done." She is a relatively old hand at canopy biology, and leads Earthwatch expeditions to Queensland's subtropical rainforests to investigate leaf damage by herbivores. Tracking leaves over time, Meg has found that while many are eaten early in their life, some manage to hang on unscathed for more than a decade.

We processed our samples at busy lab benches. Around us ethologists watched ants under microscopes, physicists calibrated sensitive instruments, chemists purified extracts from epiphytes, and physiologists collated reams of photosynthesis data.

I see now that with the advent of the raft, the canopy crane in Panama, and other innovations, the hands-on approach to tree crowns is no longer the prerogative of hardy, fervently devoted pioneers; the secret is out. Most of the participants in *Opération Canopée* had never made contact with the living rainforest roof before. On the raft, eminent professors guided young colleagues; government officials, corporate sponsors, and television broadcasters joined Hallé's crew after a few words about safety. Multiple research teams traversed the same broad sweep of canopy and shared the daunting task of charting unknown territory.

What was once a frontier enterprise was on the threshold of melding with the scientific mainstream. Having reaped the rewards of new and safer technologies, canopy researchers everywhere are increasingly spending less time fretting over how to access the treetops and more time pondering how best to record data in three-dimensional space, analyze them, and interpret the results.

EPILOGUE—DECEMBER, 1992

The rainforest canopy makes its deepest impression when faced in solitude, as most canopy explorers still face it. Alone, suspended from one of the bleached limbs of a huge *Pandanus* palm deep in the Star Mountains of New Guinea, I realize how intensely the metaphor of canopy as ocean has been entrenched in my psyche since my first climbs in the canopy. Life in the treetops has many similarities to life in the sea—with gravity adding an extra measure of difficulty. I revel in the liquid motion of foliage in tidal winds, in the grandeur of ponderous branches, in the beguiling flora and fauna that creep, glide, stretch, drift, or flap wings like fins as they move from one place to the next. I appreciate the way tropical creations seem to become more opulent in color, in form, and in lifestyle the higher I ascend. As I pause to admire a flower conceivably never witnessed by a human before, threats of biting snakes, stinging wasps, and snapping ropes fade from my mind. I'm overcome by the sensuous joy of floating within an aerial reef where trees take the place of corals.

Like other canopy biologists, I can now find peace in what once seemed an alien environment. But I wonder if this visceral empathy has been matched by scholarly understanding. Gazing at the serene canopy backdrop around me in New Guinea, I judge that the trees carry worlds on their stooped shoulders, like so many Atlases. Yet to an extent not manifested by any coral reef, the inhabitants of these worlds are stacked one upon another, so that they become submerged within the prevailing forest texture. There most of them remain, hidden from the eye and from science.

I begin to feel overwhelmed by the challenges ahead. For all our recent progress, canopy biologists have hardly begun even the most rudimentary phase of any human exploration: cataloguing what's out there. It will be an ironic legacy of this century that it has taken so long for basic rainforest surveys to get off the ground. We have a more comprehensive picture of the moon than of the rainforest, a place far more worthy of poetry and science. A tree can hold a thousand macroscopic species; perhaps eight out of ten are still without a name.

In my mind's eye, the great green bulk of one of the trees near me spins away, and these species hang in the air, forming a constellation packed with an intermingling of epiphytes, climbing plants, and animals, large and small, each one holding up its own part of the ecosystem. I marvel at the prospects of discovering usable resources among this riot of creation. Novel building materials, craft and clothing materials, fruits, vegetables, pigments, spices, and household ornamental species surely exist here. Ninety-seven kinds of rainforest epiphytes and lianas are used by one tribe alone, the Untsuri Shuar people of western Amazonia.

A quarter of all prescription drugs are derived from rainforest plants: tranquilizers, anesthetics, anticancer agents, contraceptives, and many others. Yet none of the canopy species around me and only a minute proportion of arboreal plants elsewhere on earth have been prospected for their molecular endowments. Even less well examined, tropical animals may be just as exciting biochemically: in 1992, the first known toxic bird, a species of pitahoui with poisonous skin and plumage, was identified in New Guinea. Compounds of the same chemical family, previously known only from poison dart frogs of the American tropics, have applications in medical research.

Beyond erudite biological riddles, beyond material gains, beyond the restorative potency of wilderness on the human psyche, the Midas touch of tropical rainforests may yield foundations for a more elemental wisdom. Refined over eons, with millions of species packed in a space a hundred or more feet high and extending over thousands of square miles, rainforests are the embodiment of boundless information about earth and life. Perhaps with their assistance we may even plumb the most intractable of human conundrums: Who are we? In what directions can and should our species go?

A small vanguard of scientists is only beginning to sketch out the mind-boggling intricacy of the rainforest tapestry overhead. And yet, as a consequence of these first tentative probings, there can be no doubt that the warp and weft of this tapestry is a triumph of earthly life. There is more feasting, more famine, more courtship and sex, more tender care of the young and of home, more combat and more cooperation in this arboreal realm than anyplace else on the globe. The tropical rainforest tapestry has only begun to capture our imagination.

FACING PAGE: Glass shards fleck the Himalayas and plastic sprinkles the ocean, but take us into the tropical canopy, away from the forest edge and beyond the chain saw's roar, and we will perceive no trace of the human hand (New Guinea).

ACKNOWLEDGMENTS

Around the world, arboreal ecologists have graciously shared many of their favorite trees with me. In brief:

In Costa Rica, I thank Jack Longino, who surveys ants and (recently) the general insect community of trees; Nalini Nadkarni and Teri Matelson, who share a passion for nutrient cycling and epiphytes; Pierre Berner, who studies forest dynamics and tree architecture; Jeff Luvall, who investigates the effects of atmospheric conditions on plants; Ken Clark, who puzzles over nutrients carried into the canopy by mist; and Cathy Langtimm, who looks at the adaptation of mammals to the treetops.

In Panama, I thank Alan Smith, a botanist interested broadly in physiology and canopy structure whose efforts helped make the canopy crane a reality; Gerhard Zotz,* who has been pursuing the nuances of photosythesis in a hemiepiphyte; John Tobin,* authority on the role of ants in the canopy ecosystem; and Darlyne Murawski, devoted to the population genetics of rare trees, the ingenuity of vines and the actions of pollinators. I also recognize Robin Foster, who, though he does not climb, works out the dynamics of tree communities with his colleagues, making the canopy a more comprehensible place for those of us working overhead.

Outside Central America, canopy researchers are few and far between. In South America I found Cristián Samper, who tracks the lives of hemiepiphytes and, with his associates, the movements of spectacled bears in Colombia; Terry Erwin, who collects insects in numbers almost beyond reckoning in Peru (and occasionally in other countries); Grace Savant* who compares the diets of Peruvian birds to insect availabilities; and Jay Malcolm, who appraises the survivorship of small mammals in Brazilian forest fragments.

In the Old World tropics, there are remarkably few full-time devotees of canopy research. Simmathiri Appanah in Malaysia and Savi and Nimal Gunitelleke in Sri Lanka probe into tree pollination by insects; Priya Davidar in India (who has also worked in Panama) focuses on mistletoe pollination and fruit dispersal by birds. Mark Leighton, Tim Laman, and Scott Zens in Indonesian Borneo chronicle fruit dispersal behavior by birds and mammals; Keyt Fischer in Papua New Guinea documents frugivory and foraging by marsupials. I managed to meet up with Yves Basset in the field only on the last two days of fieldwork for this book. He has worked extensively in Australia on insect community ecology, but his ongoing studies are in Papua New Guinea. I managed to see Margaret ("Meg") Lowman—who like Yves has often based her studies in Australia with ecologist Roger Kitching—in action only by carrying out a joint project on herbivory with her in Cameroon.

Francis Hallé and his associates in *Opération Canopée* I met during the canopy raft expedition they ran in Cameroon. Francis' speciality is tropical tree architecture.

I also spent time with Colin and Lauren Chapman, who use binoculars to observe arboreal monkeys and chimpanzees in Uganda. Theirs is one of the many ground-based projects on canopy vertebrates I could have visited.

I have no first-hand accounts of a few active canopy ecologists, notably Illar Muul (who studies mammals and builds walkways for ecotourism—though I experienced one of the walkways in Sabah), Maurizio Paoletti (who looks at canopy soil residents), Donald Perry and John Williams (who innovate canopy access devices), Jack Putz (who studies climbing plants), Nigel Stork (who fogs insects around the world), and Jan H. D. Wolf (who probes epiphyte communities). The list would lengthen if I included tree-climbing ecologists with projects recently begun or not now in progress; those based in subtropical or temperate zones or in non-rainforest habitats; or those focused only occasionally on the canopy.

I think it fair to say that four centers of canopy research exist today, though clearly many of the people I have mentioned belong to equally venerable institutions. The four are the University of Florida at Gainesville; the Department of Organismic and Evolutionary Biology at Harvard; the Smithsonian Tropical Research Institute (which operates the research station at Barro Colorado Island in Panama); and *Opération Canopée* in Paris (headquarters for canopy raft expeditions). Both of the universities have vigorous biology programs to which people interested in tropical canopies would naturally gravitate.

An inordinate number of canopy biologists have

*The research results of these biologists were still too preliminary to discuss in this book.

been funded by the Committee for Research and Exploration at the National Geographic Society. I've been told by committee chairman Barry C. Bishop that there has been no conscious effort by the society to promote canopy biology in particular. It's simply that National Geographic has been generally instrumental in promoting tropical study. Having received National Geographic research funds myself, I offer a joint thanks from all of us for their support.

As if that isn't enough, much of the photography for this book was completed as a *National Geographic* magazine assignment. I am delighted to acknowledge Director of Photography Thomas R. Kennedy, current editor William Graves, and former editor Wilbur E. Garrett for years of inspiration and encouragement. Also important to me have been William L. Allen, Jennifer Angle, Allen Carroll, John A. Echave, Neva L. Folk, Rick Gore (who met me before anyone else at NGS, when I weighed 135 pounds in India), David Griffin, Hillel J. Hoffmann, Kent J. Kobersteen, David Jeffery, Elizabeth A. Moize, Howard E. Paine, Constance H. Phelps, Robert M. Poole, W. Allan Royce, Susan A. Smith, Charlene S. Valeri, as well as other, first-class experts in equipment, film review, travel and layout departments. John E. Tobin and, later, Darlyne A. Murawski took time from their own research to assist with the photography of the original story produced for *National Geographic*. Robert W. Madden used his free time to help me develop the vision behind this book. Although it may seem like a long list, each person could easily merit a paragraph of their own.

I never met Dieter Plage, but greatly respected his work with *National Geographic*. A master filmmaker and naturalist, Dieter died in Sumatra on April 3, 1993, while trying to bring to life his own vision of the rainforest canopy. All lovers of the outdoors will miss him.

Thanks to friends who offered comments on most or all of the book: Peter S. Ashton, Stefan P. Cover, Deborah Fletcher, Donald and Millie Moffett, Elizabeth Royte, and the especially vigilant Darlyne A. Murawski. My editor, Maureen Graney, showed an unflagging faith in my eclectic mix of science and stories that helped turn this into more than a picture book. It's also a pleasure to thank the scientists who read individual sections concerning their areas of expertise, which in some cases required precision work with a red pen:

James D. Ackerman (epiphytes, New World)
Simmathiri Appanah (mass flowering, Asia)
Peter S. Ashton (forest dynamics, Asia)
Steven N. Austad (climbing, New Guinea)
Yves Basset (insects, Australia, New Guinea)
Kamaljit S. Bawa (tree mating systems, worldwide)
David H. Benzing (epiphytes, worldwide)
Pierre O. Berner (tree architecture, New World)
Beth Brainerd (adhesion and climbing, general)
William L. Brown, Jr. (ants, worldwide)
Colin A. Chapman (primates, Africa, New World)
Lauren J. Chapman (primates, Africa, New World)
Kenneth L. Clark (nutrient cycling, New World)
Phyllis D. Coley (herbivory, New World)
Joseph H. Connell (community dynamics, worldwide)
Stefan P. Cover (ants, New World)
Thomas B. Croat (plants, New World)
Priya Davidar (mistletoes, worldwide)
Diane W. Davidson (ant gardens, New World)
Roman Dial (lizards, insects, New World)
T. Robert Dudley (flight mechanics, general)
John F. Eisenberg (mammals, New World)
Sharon B. Emerson (gliding frogs, Asia)
Terry L. Erwin (insects, New World)
Peter Feinsinger (hummingbirds, New World)
Kathleen E. Fischer (mammals, New Guinea)
Jack B. Fisher (plant morphology, general)
Robin B. Foster (forest dynamics, New World)
Daniel Gebo (primate locomotion, general)
Alwyn H. Gentry (forest dynamics, New World)
Michael Goulding (flooded forests, New World)
Nimal Gunatilleke (pollination, Asia)
Savithri Gunatilleke (pollination, Asia)
Francis Hallé (tree architecture, worldwide)
James L. Hamrick (population genetics, New World)
Terese B. Hart (monodominant forests, Africa)
E. Allen Herre (fig wasps, New World)
Stanley R. Herwitz (hydrology, Australia)
Berthold K. Hölldobler (ants, worldwide)
Henry S. Horn (light and plant growth)
Henry F. Howe (plant dispersal, worldwide)
Robert R. Jackson (spiders, worldwide)
W. John Kress (plant reproduction)
Timothy G. Laman (frugivory, Asia)
Catherine A. Langtimm (mammals, New World)
Egbert G. Leigh, Jr. (tree architecture, New World)
Mark Leighton (frugivory, Asia)
Douglas J. Levey (birds, New World)
John T. Longino (insects, New World)
Rainer Lösch (transpiration, Africa)
Margaret D. Lowman (herbivory, Australia)
Jeffrey C. Luvall (atmospherics, New World)
Robert Mahelik (light gaps, New World)
Jay R. Malcolm (mammals, New World)
Deedra McClearn (mammals, New World)
Andrew W. Mitchell (canopy access, worldwide)
Dieter Mueller-Dombois (koa forest, Hawaii)
Darlyne A. Murawski (plant reproduction, New World)
D. H. "Paddy" Murphy (insects, Asia)

Nalini M. Nadkarni (epiphytes, New World)
Larry J. Orsak (mimicry, Old World)
Glenn D. Prestwich (termites, worldwide)
Francis E. Putz (plants, Asia, New World)
Thomas S. Ray (hemiepiphytes, New World)
Cristián Samper K. (climbers, New World)
Sarah Sargent (mistletoes, New World)
Alan P. Smith (forest strata, New World)
Margaret M. Stewart (frogs, New World)
Kyle Summers (frogs, New World)
Roger B. Swain (climbing methods, New World)
C. Richard Taylor (locomotion, general)
John W. Terborgh (forest strata, New World)
Barbara L. Thorne (termites, New World)
Merlin D. Tuttle (bats, worldwide)
Steven Vogel (fluid mechanics, general)
David B. Wake (salamanders, New World)
R. Haven Wiley (acoustics, general)
Klaus Winter (CAM plants, worldwide)
M. Scot Zens (frugivory, Asia)
Gerhard Zotz (photosynthesis, New World)

For additional insights, I'm indebted to Carol K. Augspurger, Bart Bouricius, Warren Y. Brockelman, Brian V. Brown, Diane DeSteven, Brian D. Farrell, Douglas J. Futuyma, Harry W. Greene, Pamela Hall, Robert W. Henderson, Michael Hopkins, Daniel H. Janzen, Leslie K. Johnson, Jill Landsberg, José Luis Machado, Jonathan D. Majer, Teri J. Matelson, Thomas A. McMahon, John C. Mitani, Charles A. Munn, Illar Muul, Sandra Patiño, David L. Pearson, Donald H. Pfister, J. Alan Pounds, David W. Roubik, Ira Rubinoff, Grace Serant, Hugh Spencer, Nigel E. Stork, Benito C. Tan, and Donald M. Windsor. Many other academics contributed scientific reprints.

I hope I successfully rooted out the errors these colleagues detected. I admit that in some cases a few differences of opinion—as are likely to occur in any young and vigorous science—will remain, although I try to make it clear whenever conclusions are controversial or preliminary.

I took most of the photographs over about one year of field time; gaps in my coverage of so vast a subject were inevitable. I thank canopy biologists Darlyne A. Murwaski and Timothy G. Laman for filling some of these in with their own exquisite work.

Most of all I thank Professor Edward O. Wilson at Harvard and Senior Assistant Editor Mary G. Smith at *National Geographic* for over a decade of friendship and for shaping my career in science and my career in journalism, respectively. They share an astonishingly pure sense of joy and wonder about the living world that seems to extend equally to all things, from the location of a hair on a spider's head to the grandest new scientific theory.

Canopy biologists make rewarding companions. Whereas for some scientists academics appears to be little more than the triumph of ego over boredom, these energized men and women personify for me the essence of an inquiring spirit. John Steinbeck described "what good men most biologists are" in *The Log from the Sea of Cortez*. To me, as to Steinbeck, field biologists like the ones in this book represent "the tenors of the scientific world":

> The true biologist deals with life, with teeming boisterous life, and learns something from it, learns that the first rule of life is living.

John Tobin and Ed Wilson

Gerhard Zotz

Darlyne Murawski

Terry Erwin

SELECTED REFERENCES AND NOTES

This book emphasizes ecology. For information on other rainforest issues, I recommend two books as companions to this one. For more on evolutionary processes, see Edward O. Wilson, *The Diversity of Life* (Cambridge, Mass.: Belknap Press of Harvard University Press, 1992); Wilson covers many rainforest examples. For more on conservation, see Norman Myers, *The Primary Source: Tropical Forests and Our Future* (New York: W. W. Norton, 1992).

I compress "rain forest" into one word. As Norman Myers writes, "A rainforest in Southeast Asia is so similar in structure and function to a rainforest in equatorial Africa or tropical Latin America that it has now come to be classified as a category of forest on its own: we no longer speak of 'rain forest,' but of 'rainforest.'"

A number referring to the page on which the note first applies is given at the start of each paragraph, with "L" and "R" indicating left or right column when applicable. General notes on the chapter's topic lead each chapter's section.

The opening rubrics are from William Henry Hudson, *Green Mansions* (New York: G. P. Putnam's Sons, 1904) and Alexander F. Skutch, *A Naturalist in Costa Rica* (Gainesville: University of Florida Press, 1971).

INTRODUCTION

8 Quote: Odoardo Beccari, *Wanderings in the Great Forests of Borneo*, translation (from original 1902 Italian edition) by Enrico H. Giglioli (London: Archibald Constable & Co., 1904).

8L Quote: John Fowles and Frank Horvat (photographer), *The Tree* (Boston: Little, Brown, 1979).

9L Books by the nineteenth-century explorers I mention are cited at various points in this text. Recent travel books of note include: Alexander F. Skutch, *A Naturalist Amid Tropical Splendor* (Iowa City: University of Iowa Press, 1987) and Alex Shoumatoff, *The Rivers Amazon* (London: Heinemann, 1979). Hudson's book (op. cit.) is, of course, fiction, but founded on his tropical experiences.

9L Quote: Dante Alighieri, *The Inferno*, translation by Allen Mandelbaum (Berkeley: University of California Press, 1980). Other examples of negative forest imagery: Joseph Conrad, *Heart of Darkness*, *Blackwood's Magazine* (February, March, and April 1899); Graham Greene, *Journey without Maps* (London: William Heinemann, 1936).

9R Quote: Paul W. Richards, *The Life of the Jungle* (New York: McGraw-Hill, 1970).

CHAPTER 1: *Tree Climbing for Grown-ups*

17R Quote: Richard Spruce, *Notes of a Botanist on the Amazon and Andes*, two volumes, edited by Alfred Russel Wallace (New York: Macmillan and Company, 1908).

18R Painstaking surveys of plants in an Ecuadorian rainforest were conducted by Alwyn H. Gentry and Calaway H. Dodson, "Contribution of Nontrees to Species Richness of a Tropical Rain Forest," *Biotropica* 19 (1987), 149–56. See also Alwyn H. Gentry, "Tropical Forest Biodiversity: Distributional Patterns and Their Conservational Significance," *Oikos* 63(1992), 19-28.

20L Egbert G. Leigh, Jr., scrutinizes cloud forests and elfin forests in "Tree Shape and Leaf Arrangement: A Quantitative Comparison of Montane Forests, with Emphasis on Malaysia and South India," in *Conservation in Developing Countries: Problems and Prospects* (Bombay: Oxford University Press, 1983), 119–74.

21L Barro Colorado Island, mentioned in some chapters, is on the dry extreme for rainforest; some classify it as tropical moist forest.

23L Quote: Marston Bates, *The Forest and the Sea* (New York: Vintage Books, 1960). Bates used tree crown platforms in his study of canopy mosquitoes.

24L Quote: Roman Dial, "A Food Web for a Tropical Rain Forest: The Canopy View From *Anolis*," Ph.D. dissertation, Stanford University, 1992, quoted with the permission of the author.

24R Quote: Henry Major Tomlinson, *The Sea and the Jungle* (New York: Modern Library, 1928).

25L Quote: Jacques-Yves Cousteau, *The Silent World* (New York: Time, Inc., 1950).

25R Quote: Alexander von Humboldt, *Relation Historique du Voyage aux Régions Équinoxiales du Nouveau Continent* (Paris, 1814–25).

25R Quote: William C. Beebe, G. Inness Hartley, and Paul G. Howes, *Tropical Wild Life in British Guiana* (New York: New York Zoological Society, 1917).

25R Recent views on the significance of the canopy are given by Edward O. Wilson. "Rain Forest Canopy: The High Frontier," *National Geographic* 180 (December 1991), 78–107; Margaret D. Lowman and Mark W. Moffett, "The Ecology of Tropical Rain Forest Canopies," *Trends in Ecology and Evolution* 8 (March 1993), 104–7.

CHAPTER 2: *Seeing the Forest for the Trees*

Dipterocarp trees—a group that comes up several times in this book—are discussed by Peter S. Ashton, "Dipterocarp Biology as a Window to the Understanding of Tropical Forest Structure," *Annual Review of Ecology and Systematics* 19 (1988), 347–70.

Monodominant forests are described by Terese B. Hart, John A. Hart, and Peter G. Murphy, "Monodominant and Species-Rich Forests of the Humid Tropics: Causes for their Co-Occurrence," *American Naturalist* 133 (1989), 613–33; Joseph H. Connell and Margaret D. Lowman, "Low-Diversity Tropical Rain Forests: Some Possible Mechanisms for Their Existence," *American Naturalist* 134 (1989), 88–119; Terese B. Hart, "Monospecific Dominance in Tropical Rain Forests," *Trends in Ecology and Evolution* 5 (January 1990), 6–11.

28R Quote: Carl Block, *The Head-Hunters of Borneo* (London: Sampson Low, Marston, Searle and Rivington, 1881).

30R Quote: Norman Myers, *The Primary Source: Tropical Forests and Our Future* (New York: W. W. Norton, 1992).

31R Quote: Alfred Russel Wallace, *The Malay Archipelago: The Land of the Orang-Utan and the Bird of Paradise*, two volumes (New York: Harper and Brothers, 1869).

31R Areas rich in tree species are examined by Alwyn H. Gentry, "Tree Species Richness of Upper Amazonian Forests," *Proceedings of the National Academy of Sciences* 85 (1988), 156–59. For a broad perspective, see Gentry's "Changes in Plant Community Diversity and Floristic Composition on Environmental and Geographic Gradients," *Annals of the Missouri Botanical Garden* 75 (1988), 1–34. For more on physical environment, see S. Joseph Wright, "Seasonal Drought, Soil Fertility and the Species Density of Tropical Forest

Plant Communities," *Trends in Ecology and Evolution* 7 (August 1992), 260–63.

35L The past occurrence of refugia and the value of this concept in explaining present areas of high species diversity are disputed by some scientists. See Paul A. Colinvaux, "The Past and Future Amazon," *Scientific American* 260 (May 1989), 102–8.

35L I thank Peter Ashton for information on the ongoing census of trees at least one centimeter in diameter at breast height at Lambir, Sarawak.

35R Michael Goulding, "Flooded Forests of the Amazon," *Scientific American* 266 (March 1993), 114–20.

36L Milton Lieberman and Diana Lieberman, "Patterns of Density and Dispersion of Forest Trees," in *La Selva: Ecology and Natural History of a Neotropical Rainforest*, edited by Lucinda A. McDade, Kamaljit S. Bawa, Gary S. Hartshorn, and Henry A. Hespenheide (Chicago: University of Chicago Press, in press).

36L Egbert G. Leigh, Jr., "Introduction: Why Are There So Many Kinds of Tropical Trees?" in *The Ecology of a Tropical Forest. Seasonal Rhythms and Long-Term Changes*, edited by Egbert G. Leigh, Jr., A. Stanley Rand, and Donald M. Windsor (Washington, D.C.: Smithsonian Institution Press, 1982), 63–66.

36R Pathogens and herbivores are among a number of possible "density dependent factors"—factors that cause higher mortality at higher plant densities—that could be controlling populations of competitively superior species; for example, competition between seedlings of the species for soil nutrients might produce the same effect.

37L *Platypodium* seed shadows and mortality have been documented in two papers by Carol K. Augspurger: "Seed Dispersal of the Tropical Tree, *Platypodium elegans*, and the Escape of Its Seedlings from Fungal Pathogens," *Journal of Ecology* 71 (1983), 759–71; "Offspring Recruitment Around Tropical Trees: Changes in Cohort Distance with Time," *Oikos* 40 (1983), 189–96.

37R Richard Condit, Stephen P. Hubbell, and Robin B. Foster overview methods and results from the BCI plot in "Short-Term Dynamics of a Neotropical Forest: Change Within Limits," *BioScience* 42 (1992), 822–28. Interpretations of the plot data have varied with analyses employed. Unfortunately the BCI data do not include the seedlings, where most of the density-dependent mortality occurs in species like *Platypodium*.

38L Many studies have confirmed high mortality of seeds or seedlings of common trees; see Deborah A. Clark and David B. Clark, "Spacing Dynamics of a Tropical Rain Forest Tree: Evaluation of the Janzen-Connell Model," *American Naturalist* 124 (1984), 769–88. In some cases the results have been ambiguous: Joseph H. Connell, J. G. Tracey, and Leonard J. Webb, "Compensatory Recruitment, Growth, and Mortality as Factors Maintaining Rain Forest Tree Diversity," *Ecological Monographs* 54 (1984), 141–64. The high death rate of seeds and seedlings in common species can be due either to their high densities on the forest floor or to their proximity to adult trees—though normally both factors occur in synchrony.

40L Quote: Donald R. Perry, *Life above the Jungle Floor* (New York: Simon & Schuster, 1986).

41L Julie Sloan Denslow, "Tropical Rainforest Gaps and Tree Species Diversity," *Annual Review of Ecology and Systematics* 18 (1987), 431–51. For a smorgasbord of good articles on gaps, I recommend the series of papers in *Ecology* 70, no. 3 (1989).

41R M. D. Swaine, Diana Lieberman, and Francis E. Putz cover tree longevity and growth: "The Dynamics of Tree Populations in Tropical Forest: A Review," *Journal of Tropical Ecology* 3 (1987), 359–66.

41R "Climax" was once considered a fixed endpoint to succession. Some ecologists now argue that most rainforests never reach equilibrium, at least, based on the periods of time and geographic scales that they have chosen to study.

41R The pioneer and "climax forest" tree distinction is analyzed in detail by Charles W. Welden, Steven W. Hewett, Stephen P. Hubbell, and Robin B. Foster, "Sapling Survival, Growth and Recruitment: Relationship to Canopy Height in a Neotropical Forest," *Ecology* 72 (1991), 35–50. Some people restrict the term *pioneer* only to short-lived species, thereby excluding the kapok because it survives through the whole successional process.

41R The change in light environment during succession is more complex than once thought: Alan P. Smith, Kevin P. Hogan, and Jacqueline R. Idol, "Spatial and Temporal Patterns of Light and Canopy Structure in a Lowland Tropical Moist Forest," *Biotropica* 24 (1992), 503–11; Milton Lieberman, Diana Lieberman, and Rodolfo Peralta, "Forests Are Not Just Swiss Cheese: Canopy Stereogeometry of Non-Gaps in Tropical Forests," *Ecology* 70 (1989), 550–52.

43L Deborah A. Clark, "Regeneration of Canopy Trees in Tropical Wet Forests," *Trends in Ecology and Evolution* 1 (December 1986), 150–54; Miguel Martínez-Ramos, Elena Alvarez-Buylla, José Sarukhán, and Daniel Piñero. "Treefall Age Determination and Gap Dynamics in a Tropical Forest," *Journal of Ecology* 76 (1988), 700–16.

43R In documenting the relationships between gaps and tree diversity, the characteristics of treefall gaps in monodominant versus species-rich forest is a critical subject for future study.

44R Joseph H. Connell wrote the classic paper on the "intermediate disturbance hypothesis": "Diversity in Tropical Rain Forests and Coral Reefs," *Science* 199 (1978), 1302–10. For an updated look, see Peter S. Petraitis, Roger E. Latham, and Richard A. Niesenbaum, "The Maintenance of Species Diversity by Disturbance," *Quarterly Review of Biology* 64 (1989), 393–418; also Richard J. Hobbs and Laura F. Huenneke, "Disturbance, Diversity and Invasion: Implications for Conservation," *Conservation Biology* 6 (1992), 324–37.

44R Quote: Sir Thomas Browne, *Hydriotaphia, or Urne Burial* (London: Hen. Brome, 1658).

CHAPTER 3: *A Palace of Many Floors*

One of the best thoroughgoing accounts of plant structure is Adrian D. Bell, *Plant Form: An Illustrated Guide to Flowering Plant Morphology* (New York: Oxford University Press, 1991).

50R An advanced strategy of rope climbing is described by Donald R. Perry and John Williams, "The Tropical Rain Forest Canopy: A Method Providing Total Access," *Biotropica* 13 (1981), 283–85.

51L Quote: Edred John Henry Corner, "On Thinking Big," *Phytomorphology* 17 (1967), 24–29.

51L Quote: Italo Calvino, *The Baron in the Trees*, translated by Archibald Colquhoun (London: William Collins Sons, 1959).

51R The function of buttresses remains controversial. Les Kaufman provides an intriguing point of view in "The Role of Developmental Crises in the Formation of Buttresses: A Unified Hypothesis," *Evolutionary Trends in Plants* 2 (1988), 39–51.

51R Quote: Frank M. Chapman, *My Tropical Air Castle* (New York: D. Appleton and Company, 1929).

51R Quote: Theodosius Dobzhansky, "Evolution in the Tropics," *American Scientist* 38 (1950), 209–21.

52L The scientific name of the gympie tree is *Dendrocnide excelsa*. It belongs to the Urticaceae, the nettle family.

53L Quote: Charles L. Hogue, *The Armies of the Ant* (New York: World Publishing, 1972).

53R The four basic types of leaf renewal are described by Timothy C. Whitmore, *Tropical Rain Forests of the Far East* (Oxford: Clarendon Press, 1984); other classifications have been proposed.

54L The canopy crane is depicted by Geoffrey G. Parker, Alan P. Smith, and Kevin P. Hogan, "Access to the Upper Forest Canopy with a Large Tower Crane," *BioScience* 42 (October 1992), 664–70.

54L The influence of leaves on light levels (and of light levels on leaves) is described by Henry S. Horn, *The Adaptive Geometry of Trees* (Princeton: Princeton University Press, 1971). For more detailed discussions, see Robin L. Chazdon, "Sunflecks and their Importance to Forest Understorey Plants," in *Advances in Ecological Research*, Volume 18, edited by M. Bergon, A. H. Fitter, E. David Ford, and Amyan MacFadyen (London: Academic Press, 1988), 1–63; and William K. Smith, Alan K. Knapp, and William A. Reiners, "Penumbral [light-spreading] Effects on Sunlight Penetration in Plant Communities," *Ecology* 70 (1989), 1603–9.

54R I am grateful to Pierre O. Berner for discussions of the research for his thesis, the details of which are as yet unpublished.

55L Truman P. Young, and Stephen P. Hubbell, "Crown Asymmetry, Treefalls, and Repeat Disturbance of Broad-Leaved Forest Gaps," *Ecology* 72 (1991), 1464–71.

55L Quote: Hermann Hesse, *Wandering* (New York: Farrar, Straus and Giroux, translation 1972).

55R The classic work on tree architecture is Francis Hallé, Roelof A. A. Oldeman, and P. Barry Tomlinson, *Tropical Trees and Forests: An Architectural Analysis* (Berlin: Springer-Verlag, 1978). Updates have been written by P. Barry Tomlinson, "Architecture of Tropical Plants," *Annual Review of Ecology and Systematics* 18 (1987), 1-21, and Manfred Küppers, "Ecological Significance of Aboveground Architectural Patterns in Woody Plants: A Question of Cost-Benefit Relationships." *Trends in Ecology and Evolution* 4 (December 1989), 375–79.

55R Quote: Francis S. P. Ng, "Forest Tree Biology," in *Key Environments: Malaysia*, edited by the Earl of Cranbrook (Oxford: Pergamon Press, 1988), 102–15.

58L Quote: Henry Walter Bates, *The Naturalist on the River Amazons* (London: John Murray, 1864).

58R Quote: Georg Schweinfurth, *The Heart of Africa*, translated by Ellen E. Frewer (New York: Harper and Brothers, 1874).

58R Paul W. Richards discusses whether layering is objectively demonstrable or an arbitrary subdividing of a continuum in "The Three-Dimensional Structure of Tropical Rain Forest," in *Tropical Rain Forest: Ecology and Management*, edited by Stephen L. Sutton, Timothy C. Whitmore, and A. C. Chadwick (Oxford: Blackwell Scientific Publications, 1983), 3–10.

59L Three forms of stratification may occur: stratification of species, of individual plants, and of leaf mass. Stratification could exist in any one of these ways but not in the others. Still, when young trees are excluded from consideration, the three categories probably will coincide in most cases. See discussion by Alan P. Smith, "Stratification of Temperate and Tropical Forests," *American Naturalist* 107 (1973), 671–83. Smith also explores some likely consequences of canopy layering for arboreal communities.

59R Relationships among crown shapes, light penetration, and plant stratification are explored elegantly by John Terborgh, "The Vertical Component of Plant Species Diversity in Temperate and Tropical Forests," *American Naturalist* 126 (December 1985), 760–76. Stratification increases in complexity during succession, as described by John Terborgh and Kenneth Petren, "Development of Habitat Structure Through Succession in an Amazonian Floodplain Forest," in *Habitat Structure: The Physical Arrangement of Objects in Space*, edited by Susan S. Bell, Earl D. McCoy, Henry R. Mushinsky (London: Chapman and Hall, 1991), 28–46.

62L Crown shyness has been described by Francis S. P. Ng, "Shyness in Trees," *Nature Malaysiana* 2 (1977), 34–37, and (for Costa Rican mangroves) by Francis E. Putz, Geoffrey G. Parker, and Ruth M. Archibald, "Mechanical Abrasion and Intercrown Spacing," *American Midland Naturalist* 112 (1984), 24–28. Crown shyness may be most pronounced between trees of the same species: Francis Hallé, personal communication.

63R The effects of storms on leaves is described by Steven Vogel in "Drag and Reconfiguration of Broad Leaves in High Winds," *Journal of Experimental Botany* 40 (August 1989), 941–48.

CHAPTER 4: *Gardens in the Sky*

The best treatment of epiphytes today is David H. Benzing's *Vascular Epiphytes* (New York: Cambridge University Press, 1990). Benzing's book is the source of the quote on page 73. For the most part, only references not provided by Benzing are given below.

65 Research results from the Uganda tower were presented in Alexander J. Haddow, Philip S. Corbet, and J. David Gillett, "Entomological Studies from a High Tower in Mpanga Forest, Uganda," *Transactions of the Royal Entomological Society of London* 113 (November 1961), 249–56, and subsequent articles in that issue.

68L An example of work from the Costa Rican towers: Jeffrey C. Luvall, Diana Lieberman, Milton Lieberman, Gary S. Hartshorn, and Rodolfo Peralta, "Estimation of Tropical Forest Canopy Temperatures, Thermal Response Numbers, and Evapotranspiration Using an Aircraft-based Thermal Sensor," *Photogrammetric Engineering and Remote Sensing* 56 (1990), 1393–1401.

70L Nalini M. Nadkarni has popularized her canopy root studies in "Roots that Go Out on a Limb," *Natural History*, February 1985, 43–48. She first described the ecology of canopy roots in "Canopy Roots: Convergent Evolution in Rainforest Nutrient Cycles," *Science* 214 (1981), 1023–24. The ecology of these roots has been appraised by David H. Benzing, "Aerial Roots and Their Environments," in *Plant Roots: The Hidden Half*, edited by Yoav Waisel, Amram Eshel, and Uzi Kafkafi (New York: M. Dekker, 1991), 867–85.

71L John T. Longino, "Ants Provide Substrate for Epiphytes," *Selbyana* 9 (1986), 100–3.

71R The relative unimportance of litter to soil accretion along branches in the canopy is made clear by Nalini M. Nadkarni and Teri J. Matelson, "Fine Litter Dynamics within the Tree Canopy of a Tropical Cloud Forest," *Ecology* 72 (1991), 2071–82.

72L Kenneth L. Clark and Nalini M. Nadkarni, "Nitrate and Ammonium Ions in Precipitation and Throughfall of a Neotropical Cloud Forest: Implications for Epiphyte Mineral Nutrition," *Bulletin of the Ecological Society of America* 71 (1990), 121.

72L Stanley R. Herwitz describes absorption of water and nutrients by free-dangling tree roots: "Aboveground Adventitious Roots and Stemflow Chemistry of *Ceratopetalum virchowii* in an Australian Montane Tropical Rain Forest," *Biotropica* 23 (1991), 210–18.

72L Quote: Eric Hansen, *Stranger in the Forest: On Foot Across Borneo* (Boston: Houghton Mifflin, 1988).

72R Some of Nalini M. Nadkarni's results from Olympic Peninsula rainforests are presented in "Biomass and Mineral Capital of Epiphytes in an *Acer macrophyllum* Community of a Temperate Moist Coniferous Forest, Olympic Peninsula, Washington State," *Canadian Journal of Botany* 62 (1984), 2223–28.

73L Jan H. D. Wolf, "Diversity Patterns and Biomass of Epiphytic Bryophytes and Lichens along an Altitudinal Gradient in the Northern Andes," *Annals of the Missouri Botanical Garden*, in press.

73L For the plant survey by Alwyn Gentry and Calaway Dodson, see notes on Chapter 1. Also of interest is their article, "Diversity and Biogeography of Neotropical Vascular Epiphytes," *Annals of the Missouri Botanical Garden* 74 (1987), 205–33.

77R The release of nitrogen by epiphylls is shown by Barbara L. Bentley and Edward J. Carpenter, "Direct Transfer of Newly-Fixed Nitrogen from Free-Living Epiphyllous Microorganisms to their Host Plant," *Oecologia* 63 (1984), 52–56; and Barbara L. Bentley, "Nitrogen Fixation by Epiphylls in a Tropical Rainforest," *Annals of the Missouri Botanical Garden* 74 (1987), 234–41. For another advantage of epiphylls to the host plant, see Ulrich G. Mueller and Bettina Wolf-Mueller, "Epiphyll Deterrence to the Leafcutter Ant *Atta cephalotes*," *Oecologia* 86 (1991), 36–39.

77R CAM plants store CO_2 in the form of organic acids (mainly malic acid). For a review, see Klaus Winter, "Crassulacean Acid Metabolism," in *Photosynthetic Mechanisms and the Environment*, edited by James Barber and Neil R. Baker (Amsterdam: Elsevier, 1985), 329–87. For recent canopy findings, see Klaus Winter, Gerhard Zotz, Bernhard Baur, and Karl-Josef Dietz, "Light and Dark CO_2 Fixation in *Clusia uvitana* and the Effects of Plant Water Status and CO_2 Availability," *Oecologia* 91 (1992), 47–51.

77R Charles M. Peters, Alwyn H. Gentry, and Robert O. Mendelsohn, "Valuation of an Amazonian Rainforest," *Nature* 399 (1989), 655–56.

78L Mapping epiphytes is described by Douglas Nychka and Nalini M. Nadkarni, "Spatial Analysis of Points on Tree Structures: The Distribution of Epiphytes on Tropical Trees," *University of North Carolina Institute of Statistics Mimeograph Series* no. 1971 (January 1990).

80L The relationship of epiphyte species distribution to branch diameter, inclination, and soil chemistry is discussed by Stephen W. Ingram and Nalini M. Nadkarni, "The Composition and Distribution of Epiphytic Organic Matter in a Neotropical Cloud Forest, Costa Rica" *Biotropica*, in press (1993); Peter Lesica and Robert K. Antibus, "Canopy Soils and Epiphyte Richness," *National Geographic Research and Exploration* 7 (1991), 156–65.

80R The estimate of over twenty tons of water in bromeliad tanks per acre of forest is derived from Durland Fish's article, "Phytotelmata: Flora and Fauna," in *Phytotelmata: Terrestrial Plants as Hosts for Aquatic Insect Communities*, edited by J. Howard Frank and L. Philip Lounibos (Medford, N.J.: Plexus, 1983), 1–27; the figure given is 50,000 liters per hectare.

80R The possible repercussions of epiphytes to tree communities in the New World are mentioned first by Donald R. Strong, Jr., "Epiphyte Loads, Treefalls and Perennial Forest Disruption: A Mechanism for Maintaining Higher Tree Species Richness in the Tropics without Animals," *Journal of Biogeography* 4 (1977), 215–18.

80R N. Michele Holbrook reviewed the 1991 epiphyte symposium: "Small Plants in High Places: The Conservation and Biology of Epiphytes," *Trends in Ecology and Evolution* 6 (October 1991), 314–15.

81L The interaction of rain with vegetation is discussed by Stanley R. Herwitz,"Raindrop Impact and Water Flow on the Vegetative Surfaces of Trees and the Effects on Stemflow and Throughflow Generation," *Earth Surface Processes and Landforms* 12 (1987), 425–32.

81L Canopy roughness has been documented by Geoffrey G. Parker, Alan P. Smith, and Kevin P. Hogan. "Access to the Upper Forest Canopy with a Large Tower Crane," *BioScience* 42 (October 1992), 664–70.

CHAPTER 5: *Tapping the Ground*

Much of our knowledge about climbing plants has been summarized in *The Biology of Vines*, edited by Francis E. Putz and Harold A. Mooney (New York: Cambridge University Press, 1992). For the most part, only references not provided by Putz and Mooney are given below.

84–85 All the lianas pictured are from Panama.

84L Francis E. Putz has studied rattans: "Growth Habits and Trellis Requirements of Climbing Palms (*Calamus* spp.) in North-eastern Queensland," *Australian Journal of Botany* 38 (1990), 603–8.

84L Quote: M. Raciborski, *General Plant Geography*, translated by Helen M. Massey (Warsaw: Polish Scientific Publishers, 1924).

86L Francis E. Putz recounts his tree-climbing work with lianas in "Woody Vines and Tropical Forests," *Fairchild Tropical Garden Bulletin* (October 1988), 5–13.

86R Charles R. Darwin's book is titled *The Movements and Habits of Climbing Plants* (London: John Murray, 1876).

88L Quote: Peter Matthiessen, *The Cloud Forest* (New York: Viking Penguin, 1961).

88R Quote: Alfred Russel Wallace, *Tropical Nature and other Essays* (London: Macmillan and Company, 1878).

88R The troubles lianas cause for trees are explicated by Francis E. Putz in "Lianas vs. Trees," *Biotropica* 12 (1980), 224–25. Their distribution has been studied by David B. Clark and Deborah A. Clark in "Distribution and Effects on Tree Growth of Lianas and Woody Hemiepiphytes in a Costa Rican Tropical Wet Forest," *Journal of Tropical Ecology* 6 (1990), 321–31.

89R *The Biology of Mistletoes*, edited by D. Malcolm Calder and Peter Bernhardt (New York: Academic Press, 1983).

93L In "Notes on the Natural History of Hemiepiphytes," *Selbyana* 9 (1986), 61–69, Francis E. Putz and N. Michele Holbrook describe stranglers and others. Thomas B. Croat, "Ecology and Life Forms of Araceae," *Aroideana* 11 (1988), 4–52, documents the incredible variety of growth strategies that occur in just one plant family.

94L I restrict the term *strangler* to hemiepiphytes that slowly kill their host and then become freestanding, thereby excluding large, clinging hemiepiphytes that depend on the host for lifelong support. This distinction is like that made by biologists today between "parasitoids" (which kill their host) and "parasites" (which usually don't).

94R Thomas S. Ray, Jr., describes his findings in "Slow-motion World of Plant 'Behavior' Visible in Rain Forest," *Smithsonian*, March 1979, 121–30. His paper, "Foraging Behaviour in Tropical Herbaceous Climbers," *Journal of Ecology* 80 (1992), 189–203, provides a technical updating on growth patterns.

CHAPTER 6: *Insects on a Rampage*

I have chosen to emphasize diversity. For broader reviews of insect communities, I defer to Terry L. Erwin and John E. Tobin, who are writing chapters for *Forest Canopies*, edited by Margaret D. Lowman and Nalini M. Nadkarni (Orlando, Fla.: Academic Press, in press). The results of various researchers to date have been disjointed, encompassing varied methodologies and differing results from widely separated parts of the tropics.

100L The number of described species has been tallied, with slightly differing results, by several individuals, for example, Robert M. May, "How Many Species Are There on Earth?" *Science* 241 (1988), 1441–49, from which the quote was taken.

100L Types of rarity are defined by Deborah Rabinowitz, Sara Cairns, and Theresa Dillon, "Seven Forms of Rarity and Their Frequency in the Flora of the British Isles," in *Conservation Biology: The Science of Scarcity and Diversity*, edited by Michael E. Soulé

(Sunderland, Mass.: Sinauer, 1986), 182–204; also by Kamaljit S. Bawa and Peter S. Ashton, "Conservation of Rare Trees in Tropical Rain Forests: A Genetic Perspective," in *Genetics and Conservation of Rare Plants*, edited by Donald A. Falk and Kent E. Holsinger (New York: Oxford University Press, 1991), 62–71.

100R Quote: Lewis Carroll, *Through the Looking Glass, and What Alice Found There* (London: Macmillan, 1871).

101L Terry L. Erwin describes the history of fogging in "Canopy Arthropod Biodiversity: A Chronology of Sampling Techniques and Results," *Revista Peruana de Entomología* 32 (1989), 71–77. (He wasn't the first to fog; the original attempts were in the 1960s.) Erwin found a ten-day recovery of insect abundances after fogging; Nigel E. Stork found it took longer in "The Composition of the Arthropod Fauna of Bornean Lowland Rain Forest Trees," *Journal of Tropical Ecology* 7 (1991), 161–80.

101R The estimate of the number of arthropod species was made in Terry L. Erwin, "Tropical Forests: Their Richness in Coleoptera and other Arthropod Species," *Coleopterists' Bulletin* 36 (1982), 74–75.

101R Arguments about Erwin's estimate: Kevin J. Gaston, "The Magnitude of Global Insect Species Richness," *Conservation Biology* 5 (1991), 283–96, to which Terry L. Erwin responds, "How Many Species Are There? Revisited," *Conservation Biology* 5 (1991), 330–33, and Gaston responds in turn, "Estimates of the Near-Imponderable: A Reply to Erwin," *Conservation Biology* 5 (1991), 564–66; also Nigel E. Stork, "Insect Diversity: Facts, Fiction, and Speculation," *Biological Journal of the Linnean Society* 35 (1988), 321–37.

101R Michael A. Mares, "Neotropical Mammals and the Myth of Amazonian Biodiversity," *Science* 255 (1992), 976–79.

104L The term *tourist* was first applied to insects by D. H. Murphy, "Animals in the Forest Ecosystem," in *Animal Life and Nature in Singapore*, edited by S. H. Chuang (Singapore: University of Singapore Press, 1973, 53–73). "Tourists" were first studied in temperate areas: V. C. Moran and T. R. E. Southwood, "The Guild Composition of Arthropod Communities in Trees," *Journal of Animal Ecology* 51 (1982), 289–306.

104–5 Henry F. Howe and Lynn C. Westley review herbivory in *Ecological Relationships of Plants and Animals* (New York: Oxford University Press, 1988); see also Phyllis D. Coley and T. Mitchell Aide, "Comparison of Herbivory and Plant Defenses in Temperate and Tropical Broad-Leaved Forests," in *Plant-Animal Interactions: Evolutionary Ecology in Tropical and Temperate Regions*, edited by Peter W. Price, Thomas M. Lewinsohn, G. Wilson Fernandes, and Woodruff W. Benson (New York: John Wiley & Sons, 1991), 25–49.

104R Size and metabolism of vertebrate leaf-feeders in the canopy are assessed by John F. Eisenberg, "The Evolution of Arboreal Herbivores in the Class Mammalia," and T. Bauchop, "Digestion of Leaves in Vertebrate Arboreal Folivores," in *The Ecology of Arboreal Folivores*, edited by G. Gene Montgomery (Washington, D.C.: Smithsonian Institution Press, 1978), 135–52, 193–204, respectively.

105L Daniel H. Janzen discusses the efficacy of latex (rubber) in "Plant Defenses against Animals in the Amazonian Rainforest," in *Key Environments: Amazonia*, edited by Ghillean T. Prance and Thomas E. Lovejoy (Oxford: Pergamon Press, 1985), 207–17.

105L Gympie trees are actually heavily attacked: Margaret D. Lowman, "Leaf Growth Dynamics and Herbivory in Five Species of Australian Rain-forest Canopy Trees," *Journal of Ecology*, 80 (1992), 433–47.

105L Although young leaves are generally the favored foods, the colorful fresh leaf flushes in some trees at first contain few nutrients, making them less tempting; see Thomas A. Kursar and Phyllis D. Coley, "Delayed Greening in Tropical Leaves: An Anti-Herbivore Defense?" *Biotropica* 24 (1992), 256–62.

105R Quote: Henry Morton Stanley, *In Darkest Africa, or the Quest, Rescue, and Retreat of Emin Governor of Equatoria* (New York: Charles Scribner's Sons, 1891). What Stanley saw were presumably sulfur butterflies, not moths.

105R Associations of herbivorous beetles (chrysomelids) and predatory ones (staphylinids) with trees are shown by Brian D. Farrell and Terry L. Erwin, "Leaf-Beetle Community Structure in an Amazonian Rainforest Canopy," in *Biology of Chrysomelidae*, edited by Pierre Jolivet, E. Petitpierre, and Ting H. Hsiao (Dordrecht: Kluwer Academic Publishers, 1988), 73–90. See also Brian D. Farrell, Charles Mitter, and Douglas J. Futuyma, "Diversification at the Insect-Plant Interface," *BioScience* 42 (January 1992), 34–42.

105R Yves Basset has written several detailed articles on herbivore associations, especially in the subtropics, for example, "Host Specificity of Arboreal and Free-Living Insect Herbivores in Rain Forests," *Biological Journal of the Linnean Society* 47 (1992), 115–33.

107L Nalini M. Nadkarni and John T. Longino, "Invertebrates in Canopy and Ground Organic Matter in a Neotropical Montane Forest, Costa Rica," *Biotropica* 22 (1990), 286–89; also Maurizio G. Paoletti, Robin A. J. Taylor, Benjamin R. Stinner, Deborah H. Stinner, and David H. Benzing, "Diversity of Soil Fauna in the Canopy and Forest Floor of a Venezuelan Cloud Forest," *Journal of Tropical Ecology* 7 (1991), 373–83. For most invertebrates, the densities are often higher in ground soils, presumably because of the harsh conditions in the upper canopy (see Chapter 4).

107R Quote: Michael Goulding, *Amazon: The Flooded Forest* (New York: Sterling Publishing Company, 1990). For examples of such migrations, see Joachim Adis, "Seasonal *Igapó*-Forests of Central Amazonia Blackwater Rivers and Their Terrestrial Arthropod Fauna," in *The Amazon*, edited by Harold Sioli (Dordrecht: Dr. W. Junk, 1984), 245–68.

109L Henk Wolda's light-trap data are analyzed in "Trends in Abundance of Tropical Forest Insects," *Oecologia* 89 (1992), 47–52.

109L Quote: Thomas Belt, *The Naturalist in Nicaragua* (London: John Murray, 1874).

109L Bromeliad occupants are described by Albert M. Laessle, "A Micro-Limnological Study of Jamaican Bromeliads," *Ecology* 42 (1961), 499–517; J. Howard Frank, "Bromeliad Phytotelmata and their Biota, especially mosquitos," in *Phytotelmata: Terrestrial Plants as Hosts for Aquatic Insect Communities*, edited by J. Howard Frank and L. Philip Lounibos (Medford, N. J.: Plexus, 1983), 101–28.

109L The Jamaican crab's behavior is described by Rudolf Diesel, "Parental Care in an Unusual Environment: *Metopaulias depressus* (Decapoda: Grapsidae), a crab that lives in epiphytic bromeliads," *Animal Behaviour* 38 (1989), 561–75, and "Maternal Care in the Bromeliad Crab, *Metopaulias depressus:* Protection of Larvae from Predation by Damselfly Nymphs," *Animal Behaviour* 43 (1992), 803–12.

109L The use of bromeliads by poison dart frogs was documented first by Peter Weygoldt in "Complex Brood Care and Reproductive Behavior in Captive Poison-Arrow Frogs, *Dendrobates pumilio* O. Schmidt," *Behavioral Ecology and Sociobiology* 7 (1980), 329–32, and since then confirmed in the canopy by Kyle Summers (currently at Queens University in Ontario) and other herpetologists. A discussion by Mark W. Moffett is in press in *National Geographic Magazine.*

109R Carnivory in some bromeliads needs documentation, though convincing work has been carried out on a terrestrial species by Thomas J. Givnish, Elizabeth L. Burkhardt, Ruth E. Happel, and Jason D. Weintraub, "Carnivory in the Bromeliad *Brocchinia reducta* with a Cost/Benefit Model for the General Restriction of Carnivorous Plants to Sunny, Moist, Nutrient-Poor Habitats," *American Naturalist* 124 (1984), 479–97.

110L Quote: Sir John Bowring, *The Kingdom and People of Siam; With a Narrative of the Mission to that Country in 1855*, two volumes (London: John W. Parker and Son, 1857).

111R Some of Terry Erwin's ant samples were analyzed by Edward O. Wilson, "The Arboreal Ant Fauna of Peruvian Amazon Forests: A First Assessment," *Biotropica* 19 (1987), 245–51.

111R Quote: H. Elliott McClure, "The Secret Life of a Tree," *Animal Kingdom* 80 (1977), 16–23.

112L Quote: William L. Brown, Jr., "A Comparison of the Hylean and Congo-West African Rain Forest Ant Faunas," in *Tropical Forest Ecosystems in Africa and South America: A Comparative Review*, edited by Betty J. Meggers, Edward S. Ayensu, and W. Donald Duckworth (Washington: Smithsonian Institution Press, 1973), 161–85 .

112L The decline of ants in the canopy with forest altitude has been shown by Nigel E. Stork and Martin J. D. Brendell, "Variation in the Insect Fauna of Sulawesi Trees with Season, Altitude and Forest Type," in *Insects and the Rain Forests of South East Asia (Wallacea)*, edited by W. J. Knight and J. D. Holloway (London: Royal Entomological Society of London, 1990), 173–90.

112L John T. Longino and Nalini M. Nadkarni, "A Comparison of Ground and Canopy Leaf Litter Ants (Hymenoptera: Formicidae) in a Neotropical Montane Forest," *Psyche* 97 (1990), 81–93.

112–15 Bert Hölldobler and Edward O. Wilson, *The Ants* (Cambridge, Mass.: Belknap Press of Harvard University Press, 1990). They exhaustively survey the literature on ants, including leaf cutters, weaver ants, and ant-plant interactions.

112R Mark W. Moffett, "Cooperative Foraging in *Daceton*, with a Survey of Group Transport in Ants," *National Geographic Research and Exploration* 8 (1992), 220–31; Bert Hölldobler, Jackie Palmer, and Mark W. Moffett, "Chemical Communication in the Dacetine Ant *Daceton armigerum*," *Journal of Chemical Ecology* 16 (1990), 1207–20.

112R Ant-plant coevolution has been covered recently in *Ant-Plant Interactions*, edited by Camilla R. Huxley and David F. Cutler (New York: Oxford University Press, 1991).

113L Diane W. Davidson and William W. Epstein describe ant gardens in "Epiphytic Associations with Ants," in *Vascular Plants as Epiphytes*, edited by Ulrich Lüttge (New York: Springer-Verlag, 1989), 200–33; see also Jimmy L. Seidel, William W. Epstein, and Diane W. Davidson, "Neotropical Ant Gardens. 1. Chemical Constituents," *Journal of Chemical Ecology* 16 (1990), 1791–1816.

113R Mite-plant symbioses are treated by Dennis J. O'Dowd and Mary F. Willson, "Associations between Mites and Leaf Domatia," *Trends in Ecology and Evolution* 6 (June 1991), 179–82.

114L The territorial subdividing of the canopy by ants has been documented by Jonathan D. Majer and P. Camer-Pesci, "Ant Species in Tropical Australian Tree Crops and Native Ecosystems—Is there a Mosaic?" *Biotropica* 23 (1991), 173–81. For mangroves, see Eldridge S. Adams, "Territory defense by the Ant *Azteca trigona*: Maintenance of an Arboreal Ant Mosaic," unpublished manuscript.

114L To be accurate, for substrate-bound creatures like ant workers, the canopy is more than two-dimensional, but less than three. For foraging rules on two-dimensional leaf sprays, read Rudolph Jander, "Arboreal Search in Ants: Search in Branches (Hymenoptera: Formicidae)," *Journal of Insect Behavior* 3 (1990), 515–27.

114R Bullet ant behavior is elucidated by Michael D. Breed, Jennifer H. Fewell, A. J. Moore and Kristina R. Williams, "Graded recruitment in a ponerine ant," *Behavioral Ecology and Sociobiology* 20 (1987), 407–11; and by Michael D. Breed and Jon Harrison, "Individually Discriminable Recruitment Trails in a Ponerine Ant," *Insectes Sociaux* 34 (1987), 222–26.

114R Lilian U. Gadrinab and Monique Belin, "Biology of the Green Spots in the Leaves of some Dipterocarps," *Malaysian Forester* 44 (1981), 253–66.

115R Phyllis D. Coley and T. Mitchell Aide, "Red Coloration of Tropical Young Leaves: A Possible Antifungal Defence?" *Journal of Tropical Ecology* 5 (1989), 293–300.

CHAPTER 7: *Furred and Feathered on Top of the World*

There is extensive literature for the topics in this chapter, at least for primates. Climbing, leaping, gliding, and flying are described elegantly by Steven Vogel, *Life's Devices: The Physical World of Animals and Plants* (Princeton: Princeton University Press, 1988); Thomas A. McMahon and John Tyler Bonner, *On Size and Life* (New York: Scientific American Books, 1983); R. McNeill Alexander, *Exploring Biomechanics* (New York: Scientific American Books, 1992); Milton Hildebrand, Dennis M. Bramble, Karel F. Liem, and David B. Wake, editors, *Functional Vertebrate Morphology* (Cambridge, Mass.: Belknap Press of Harvard University Press, 1985); Peter S. Rodman and John G.H. Cant, editors, *Adaptations for Foraging in Nonhuman Primates* (New York: Columbia University Press, 1984).

117 The spectacled bear mother I describe had been gradually released into the wild after being reared in captivity. Bernard Peyton, "Ecology, Distribution, and Food Habits of Spectacled Bears, *Tremarctos ornatus*, in Peru." *Journal of Mammalogy* 61 (1980), 639–52. Jorgé Enrique Orejuela initiated studies on the bears at La Planada; I thank Cristián Samper for his assistance.

118L Pierre Charles-Dominique, *Ecology and Behaviour of Nocturnal Primates* (New York: Columbia University Press, 1977).

119R Rainforest trees with hollow cores supply the equivalent of caves for arboreal animals. Insulated by wood and relatively sequestered from air currents, they are tranquil retreats for many species that might otherwise not survive in harsh canopy conditions. Hollow trees are healthy; in fact, the rain of guano and corpses within them can be exploited as compost by roots growing into the tree's core. See T. A. Dickinson and E. V. J. Tanner, "Exploitation of Hollow Trunks by Tropical Trees," *Biotropica* 10 (1978), 231–33.

121L Quote: Ivan T. Sanderson, *Animal Treasure* (New York: Viking Press, 1937).

122L Quote: Andrew W. Mitchell, *The Enchanted Canopy* (New York: Macmillan Publishing Company, 1986). Mitchell covers many of the life histories of the vertebrates mentioned in this chapter.

122L Quote: Thomas Belt, *The Naturalist in Nicaragua* (London: John Murray, 1874).

122R Roman Dial, "A Food Web for a Tropical Rain Forest: The Canopy View From *Anolis*," Ph.D. dissertation, Stanford University, 1992.

123L Cathy Langtimm's and John Endler's results on coloration are examined by Richard A. Kiltie, "New Light on Forest Shade," *Trends in Ecology and Evolution* 8 (February 1993), 39–40.

123R R. Haven Wiley and Douglas G. Richards describe vocalizations in forest settings: "Adaptations for Acoustic Communication in Birds: Sound Transmission and Signal Detection," in *Acoustic Communication in Birds*, edited by Donald E. Kroodsma, Edward H. Miller, and Henri Ouellet (New York: Academic Press, 1982), 131–81.

124R Absorbing ideas about the importance of catching prey, reaching food at branch tips, and other matters in the origin of adaptations in arboreal primates are critiqued by D. Tab Rasmussen,

"Primate Origins: Lessons from a Neotropical Marsupial," *American Journal of Primatology* 22 (1990), 263–77; and by Robert W. Sussman, "Primate Origins and the Evolution of Angiosperms," *American Journal of Primatology* 23 (1991), 209–23.

125L Quote: Charles Waterton, *Wanderings in South America* (London: J. Mawman, 1825).

125L John G. Fleagle, *Primate Adaptation and Evolution* (San Diego: Academic Press, 1988); Deedra McClearn, "Locomotion, Posture, and Feeding Behavior of Kinkajous, Coatis, and Raccoons," *Journal of Mammalogy* 73 (1992), 245–61.

125R Matt Cartmill, "Pads and Claws in Arboreal Locomotion," in *Primate Locomotion*, edited by Farish A. Jenkins, Jr. (New York: Academic Press, 1974), 45–83.

125R David B. Wake, "Adaptive Radiation of Salamanders in Middle American Cloud Forests," *Annals of the Missouri Botanical Garden* 74 (1987), 242–63.

126L Gavin Hanna and W. Jon P. Barnes, "Adhesion and Detachment of the Toe Pads of Tree Frogs," *Journal of Experimental Biology* 155 (1991), 103–25.

126L Ernest E. Williams and Jane A. Peterson, "Convergent and Alternative Designs in the Digital Adhesive Pads of Scincid Lizards," *Science* 215 (1982), 1509–11; Anthony P. Russell, "A Contribution to the Functional Analysis of the Foot of the Tokay, *Gekko gecko* (Reptilia: Gekkonidae)," *Journal of Zoology, London* 176 (1975), 437–76.

126L Elizabeth Procter-Gray, "Kangaroos up a Tree," *Natural History*, January 1990, 61–66.

126R G. Gene Montgomery and Mel E. Sunquist, "Habitat Selection and Use by Two-Toed and Three-Toed Sloths," in *The Ecology of Arboreal Folivores*, edited by G. Gene Montgomery (Washington, D.C.: Smithsonian Institution Press, 1978), 329–59.

128L The chameleon tongue is extremely complex: Peter C. Wainwright and Albert F. Bennett, "The Mechanism of Tongue Projection in Chameleons," *Journal of Experimental Biology* 168 (1992), 1–40.

128L Harry Greene, "Diet and Arboreality in the Emerald Monitor, *Varanus prasinus*, with Comments on the Study of Adaptation," *Fieldiana: Zoology* 31 (1986), 1–12.

128L Tropical arboreal snakes form a distinct morphological class: see Craig Guyer and Maureen A. Donnelly, "Length-Mass Relationships among an Assemblage of Tropical Snakes in Costa Rica," *Journal of Tropical Ecology* 6 (1990), 65–76; Harvey B. Lillywhite and Robert W. Henderson, "Behavioral and Functional Ecology of Arboreal Snakes," in *Snakes: Ecology and Behavior*, edited by Richard Seigel and Joseph T. Collins (New York: McGraw-Hill, 1993), 1–48 .

128–29 I thank the mammologists mentioned in Chapter 7 for discussions of mammal sizes at their sites. John F. Eisenberg has shown that small leaf-eating mammals are scarce in the canopy in "The Evolution of Arboreal Herbivores in the Class Mammalia," in *The Ecology of Arboreal Folivores*, edited by G. Gene Montgomery (Washington, D.C.: Smithsonian Institution Press, 1978), 135–52 ; it seems likely this finding can be extended to mammals generally. Small vertebrates may also avoid the upper canopy because of high levels of attrition caused by birds of prey.

128–29 In writing this section, I found the books named at the start of the notes for Chapter 7 helpful, as were several other publications: John G. H. Cant, "Positional Behavior and Body Size of Arboreal Primates: A Theoretical Framework of Field Studies and an Illustration of its Application," *American Journal of Physical Anthropology* 88 (1992), 273–83; Françoise K. Jouffroy, M. Holly Stack, and Carsten Niemitz, editors, *Gravity, Posture and Locomotion in Primates* (Florence: Editrice "Il Sedicesimo," 1990); and J. Alan Pounds, "Habitat Structure and Morphological Patterns in Arboreal Vertebrates," in *Habitat Structure: The Physical Arrangement of Objects in Space*, edited by Susan S. Bell, Earl D. McCoy, and Henry R. Mushinsky (London: Chapman and Hall, 1991), 109–19.

130L Quote: Robert W. Shelford, *A Naturalist in Borneo* (London: T. Fisher Unwin, 1916).

130L James A. Oliver, " 'Gliding' in Amphibians and Reptiles, with a Remark on an Arboreal Adaptation in the Lizard, *Anolis carolinensis carolinensis* Voigt," *American Naturalist* 85 (1951), 171–76.

130R Margaret M. Stewart, "Frequent Fliers," *Natural History*, February 1992, 43–49.

130R Sharon B. Emerson and Mimi A. R. Koehl, "The Interaction of Behavioral and Morphological Change in the Evolution of a Novel Locomotor Type: 'Flying' Frogs," *Evolution* 44 (1990), 1931–46.

131L Quote: Karl P. Schmidt and Robert F. Inger, *Living Reptiles of the World* (Garden City, N. Y.: Doubleday, 1957).

131 Louise H. Emmons and Alwyn H. Gentry, "Tropical Forest Structure and the Distribution of Gliding and Prehensile-Tailed Vertebrates," *American Naturalist* 121 (1983), 513–24; Robert Dudley and Phil DeVries, "Tropical Rain Forest Structure and the Geographical Distribution of Gliding Vertebrates," *Biotropica* 22 (1990), 432–34.

132L Quote: Archie Carr, *High Jungles and Low* (Gainesville: University of Florida Press, 1953).

132L U. M. Norberg and J. M. V. Rayner, "Ecological Morphology and Flight in Bats (Mammalia; Chiroptera): Wing Adaptations, Flight Performance, Foraging Strategy and Echolocation," *Philosophical Transactions of the Royal Society of London, B* 316 (1987), 335–427.

132R Some bird foraging tactics: James V. Remsen, Jr., and Theodore A. Parker III, "Arboreal Dead-Leaf-Searching Birds of the Neotropics," *The Condor* 86 (1984), 36–41; Russell Greenberg and Judy Gradwohl, "Leaf Surface Specializations of Birds and Arthropods in a Panamanian Forest," *Oecologia* 46 (1980), 115–24; Russell Greenberg, "Dissimilar Bill Shapes in New World Tropical Versus Temperate Forest Foliage-Gleaning Birds," *Oecologia* 49 (1981), 143–47.

133L Jay R. Malcolm, "Comparative Abundances of Neotropical Small Mammals by Trap Height," *Journal of Mammalogy* 72 (1991), 188–92.

133L Richard O. Bierregaard, Jr., and Thomas E. Lovejoy, "Effects of Forest Fragmentation on Amazonian Understory Bird Communities," *Acta Amazonica* 19 (1989), 215–41.

133R Vertebrates and rainforest succession are discussed by Jay R. Malcolm (unpublished thesis, University of Florida), and by Douglas J. Levey, "Tropical Wet Forest Treefall Gaps and Distributions of Understory Birds and Plants," *Ecology* 69 (1988), 1076–89.

133 Stratification in birds was studied by Bette A. Loiselle, "Bird Abundance and Seasonality in a Costa Rican Lowland Forest Canopy," *The Condor* 90 (1988), 761–72; Charles A. Munn, "Permanent Canopy and Understory Flocks in Amazonia: Species Composition and Population Density," in *Neotropical Ornithology*, edited by P. A. Buckley, Mercedes S. Foster, Eugene S. Morton, Robert S. Ridgely, and Francine G. Buckley (Washington D.C.: Ornithological Monographs No. 36, American Ornithologists' Union, 1985), 683–712. Charles A. Munn decribes his canopytop experiences in "The Real Macaws," *Animal Kingdom*, September 1988, 20–33.

CHAPTER 8: *A Floral Symphony*

For a general treatment, see Knut Faegri and Leendert van der Pijl, *The Principles of Pollination Ecology*, 3rd revised edition (Oxford: Pergamon Press, 1978); also *Pollination Biology*, edited by Leslie Real (Orlando, Fla.: Academic Press, 1983), and general references for Chapter 9.

138R Simmathiri Appanah and H. T. Chan write of the behavior of pollinating thrips in "Thrips: The Pollinators of Some Dipterocarps," *Malaysian Forester* 44 (1981), 234–52.

139R Asia's mass flowering is described by Simmathiri Appanah, "General Flowering in the Climax Rain Forests of South-East Asia," *Journal of Tropical Ecology* 1 (1985), 225–40.

143L Temperature as a cue in mass flowering is revealed by Peter S. Ashton, Thomas J. Givnish, and Simmathiri Appanah, "Staggered Flowering in the Dipterocarpaceae: New Insights into Floral Induction and the Evolution of Mast Fruiting in the Aseasonal Tropics," *American Naturalist* 132 (1988), 44–66. These authors also review the seasonal-climate origin of the dipterocarps.

143R The varied ways birds take advantage of epiphytes are considered by Nalini M. Nadkarni and Teri J. Matelson, "Bird Use of Epiphyte Resources in Neotropical Trees," *The Condor* 91 (1989), 891–907.

143R Sexual mimicry and other forms of pollinator deceit are among the topics covered by James D. Ackerman, "Coping with the Epiphytic Existence: Pollination Strategies," *Selbyana* 9 (1986), 52–60; and by L. Anders Nilsson, "Orchid Pollination Biology," *Trends in Ecology and Evolution* 7 (1992), 255–59.

143R Orchid bees are under study by James D. Ackerman. See, for example, his "Geographic and Seasonal Variation in Fragrance Choices and Preferences of Male Euglossine Bees," *Biotropica* 21 (1989), 340–47.

145L The relationship between pollinator type and tree-species richness has been described by Simmathiri Appanah, "Plant-Pollinator Interactions in Malaysian Rain Forests," in *Reproductive Ecology of Tropical Forest Plants,* edited by Kamaljit S. Bawa and Malcolm Hadley (Paris: UNESCO, 1990), 85–101.

145R Judith L. Bronstein, "Seed Predators as Mutualists: Ecology and Evolution of the Fig/Pollinator Interaction," in *Insect-Plant Interactions,* Volume IV, edited by Elizabeth Bernays (Boca Raton: CRC Press, 1992), 1–44. For a new development, see Edward Allen Herre, "Population Structure and the Evolution of Virulence in Nematode Parasites of Fig Wasps," *Science* 259 (1993), 1442–45.

145R Under unusual conditions, the one-to-one relationship between figs and their pollinators can be imperfect; see Anthony B. Ware and Stephen G. Compton, "Breakdown of Pollinator Specificity in an African Fig Tree," *Biotropica* 24 (1992), 544–49.

146R Pollination biology of Sri Lankan dipterocarps is described by S. Dayanandan, D. N. C. Attygalla, A. W. W. L. Abeygunasekera, I. A. U. N. Gunatilleke, and C. V. S. Gunatilleke, "Phenology and Floral Morphology in Relation to Pollination of Some Sri Lankan Dipterocarps," in *Reproductive Ecology of Tropical Forest Plants,* edited by Kamaljit S. Bawa and Malcolm Hadley (Paris: UNESCO, 1990).

146R W. John Kress and James H. Beach assess stratification of pollinators in different parts of the globe, "Flowering Plant Reproductive Systems at La Selva Biological Station," in *La Selva: Ecology and Natural History of a Neotropical Rainforest,* edited by Lucinda A. McDade, Kamaljit S. Bawa, Gary S. Hartshorn, and Henry A. Hespenheide (Chicago: University of Chicago Press, in press).

147R Darlyne A. Murawski appraises *Heliconius* behavior in "Floral Resource Variation, Pollinator Response, and Potential Pollen Flow in *Psiguria warscewiczii.*" *Ecology* 68 (1987), 1273–82.

148L Quote: Henry N. Ridley, *The Dispersal of Plants throughout the World* (Ashford, Kent: L. Reeve and Company, 1930).

148R Quote: Frank G. Browne, *Forest Trees of Sarawak and Brunei and Their Products* (Kuching: Government Printing Office, 1954).

150L Effects of a continuous emergent canopy on the main canopy is described in a richly detailed paper by Peter S. Ashton and Pamela Hall, "Comparisons of Structure among Mixed Dipterocarp Forest of North-Western Borneo," *Journal of Ecology* 80 (1992), 459–81.

150R The paucity of Asian animals, the migratory behavior of seed predators, and other issues are explored by Daniel H. Janzen, "Tropical Blackwater Rivers, Animals, and Mast Fruiting by the Dipterocarpaceae," *Biotropica* 6 (1974), 69–103, and by Simmathiri Appanah, "General Flowering in the Climax Rain Forests of South-East Asia," *Journal of Tropical Ecology* 1 (1985), 225–40.

150R Stan L. Lindstedt and Mark S. Boyce, "Seasonality, Fasting Endurance, and Body Size in Mammals," *American Naturalist* 125 (1985), 873–78. For another point of view, see Cris Cristoffer, "Body Size Differences Between New World and Old World, Arboreal, Tropical Vertebrates: Cause and Consequences," *Journal of Biogeography* 14 (1987), 165–72.

CHAPTER 9: *Treetop Games between Plants and Animals*

Many issues in Chapter 9 are deftly examined in *Seed Dispersal,* edited by David R. Murray (New York: Academic Press, 1986); Henry F. Howe and Lynn C. Westley, *Ecological Relationships of Plants and Animals* (New York: Oxford University Press, 1988); Kamaljit S. Bawa, "Plant-Pollinator Interactions in Tropical Rain Forests," *Annual Review of Ecology and Systematics* 21 (1990), 399–422. I thank Mark Leighton, Scott Zens, and Timothy G. Laman for enlightening conversations about their unpublished results on Bornean seed dispersal syndromes and hemiepiphytic figs.

154L Splendid adventures with orangutans, observed from both canopy and ground levels, are available in Barbara Harrisson's *Orang-Utan* (London: William Collins Sons, 1962).

154R Fruit—technically any structure bearing or containing seeds—need not be fleshy and edible, though for ease of discussion I apply the term only to edible forms in the text.

154R Many frugivores, including the orangutan, are dispersers of some plant species and seed predators of others.

155L Fakhri A. Bazzaz contemplates the world of plants in terms more commonly reserved for animals in "Habitat Selection in Plants," *American Naturalist* 137 supplement (1991), 116–30.

156L James L. Hamrick and Darlyne A. Murawski, "The Breeding Structure of Tropical Tree Populations," *Plant Species Biology* 5 (1990), 157–65; James L. Hamrick and Darlyne A. Murawski, "Levels of Allozyme Diversity in Populations of Uncommon Neotropical Tree Species," *Journal of Tropical Ecology* 7 (1991), 395–99; Darlyne A. Murawski and James L. Hamrick, "The effect of the Density of Flowering Individuals on the Mating Systems of Nine Tropical Tree Species," *Heredity* 67 (1991), 167–74; James L. Hamrick, Darlyne A. Murawski, and John D. Nason, "The Influence of Seed Dispersal Mechanisms on the Genetic Structure of Tropical Tree Populations," *Vegetatio* (1993, in press).

156R Cuipo is investigated by Darlyne A. Murawski and James L. Hamrick, "The Mating System of *Cavanillesia platanifolia* under Extremes of Flowering-Tree Density: A Test of Predictions," *Biotropica* 24 (1992), 99–101.

157L Quote: Alex Shoumatoff, *In Southern Light: Trekking Through Zaire and the Amazon* (New York: Simon & Schuster, 1986).

157L A generalized small-insect flowering syndrome is defined, for example, by Kamaljit S. Bawa, Stephen H. Bullock, Donald R. Perry, R. E. Coville, and M. H. Grayum, "Reproductive Biology of Tropical Lowland Rain Forest Trees. II. Pollination Systems," *American Journal of Botany* 72 (1985), 346–56.

157R Kamaljit S. Bawa, "Evolution of Dioecy in Flowering Plants," *Annual Review of Ecology and Systematics* 11 (1980), 15–39.

158L L. Anders Nilsson, Lars Jonsson, Lydia Ralison, and Emile Randrianjohany, "Angraecoid Orchids and Hawkmoths in Central Madagascar: Specialized Pollination Systems and Generalist Foragers," *Biotropica* 19 (1987), 310–18.

158R Basic ideas about dispersal are raised by Nathaniel T. Wheelwright and Gordon H. Orians, "Seed Dispersal by Animals: Contrasts with Pollen Dispersal, Problems of Terminology and Constraints on Coevolution," *American Naturalist* 119 (1982), 402–13.

159L Carol K. Augspurger considers wind dispersal in "Morphology and Dispersal Potential of Wind-Dispersed Diaspores of Neotropical Trees," *American Journal of Botany* 73 (1986), 353–63.

159R Theodore H. Fleming, Randall Breitwisch, and George H. Whitesides, "Patterns of Tropical Vertebrate Frugivore Diversity," *Annual Review of Ecology and Systematics* 18 (1987), 91–109.

159R Flightless pollinators are appraised by Robert W. Sussman and Peter H. Raven, "Pollination by Lemurs and Marsupials: An Archaic Coevolutionary System," *Science* 200 (1978), 731–36.

159R Michael Goulding describes fruit dispersal by fish in *The Fishes and the Forest* (Berkeley: University of California Press, 1980).

160L Larry L. Rockwood, "Seed Weight as a Function of Life Form, Elevation and Life Zone in Neotropical Forests," *Biotropica* 17 (1985), 32–39; Susan A. Foster and Charles H. Janson, "The Relationship Between Seed Size and Establishment Conditions in Tropical Woody Plants," *Ecology* 66 (1985), 773–80.

160L Chapter 2 reports on mortality of offspring near the parent tree.

161L Claude Martin reviews seed dispersal by elephants and other topics in *The Rainforests of West Africa*, translated by Linda Tsardakas (Boston: Birkhäuser-Verlag, 1990).

161L Daniel H. Janzen and P. Martin, "Neotropical Anachronisms: What the Gomphotheres Ate," *Science* 215 (1982), 19–27; but see Henry F. Howe, "Gomphothere Fruits: A Critique," *American Naturalist* 125 (1985), 853–65.

161L Oilbirds are discussed by David W. Snow in *The Web of Adaptation: Bird Studies in the American Tropics* (New York: Quadrangle, 1976).

161R Priya Davidar, "Fruit Structure in two Neotropical Mistletoes and its Consequences for Seed Dispersal," *Biotropica* 19 (1987), 137–39; Priya Davidar, "Similarity between Flowers and Fruits in some Flowerpecker Pollinated Mistletoes," *Biotropica* 15 (1983), 32–37.

162L Mark Leighton, "Modeling Diet Selectivity by Bornean Orangutans: Evidence for Integration of Multiple Criteria in Fruit Selection," *International Journal of Primatology* 14 (1993), 1–57.

162R Quote: Eric Mjöberg, *Forest Life and Adventures in the Malay Archipelago*, English translation (from original 1928 Swedish edition) by A. Barwell (London: George Allen and Unwin, 1930).

163L Specialized frugivores include oilbirds (see above) and certain birds of paradise: Bruce M. Beehler, *A Naturalist in New Guinea* (Austin: University of Texas Press, 1991). In less extreme instances, tropical birds take fruit as big as their bills can manage; wide-gaped birds have the option of eating either large or small fruits: Nathaniel T. Wheelwright, "Fruit Size, Gape Width, and the Diets of Fruit Eating Birds," *Ecology* 66 (1985), 808–18.

163L The role of one fruit trait in dispersal is discussed by Mary F. Willson and Christopher J. Whelan, "The Evolution of Fruit Color in Fleshy-Fruited Plants," *American Naturalist* 136 (1990), 790–809.

164L For a critique of fruit syndromes, see Kathleen E. Fischer and Colin A. Chapman, "Frugivores and Fruit Syndromes: Differences in Patterns at the Genus and Species Level," *Oikos*, in press.

164L Richard Primack, "Relationships among Flowers, Fruits and Seeds," *Annual Review of Ecology and Systematics* 18 (1987), 409–30.

164R Quote: Robin B. Foster, "Famine on Barro Colorado Island," in *The Ecology of a Tropical Forest. Seasonal Rhythms and Long-Term Changes*, edited by Egbert G. Leigh, Jr., A. Stanley Rand, and Donald M. Windsor (Washington, D.C.: Smithsonian Institution Press, 1982), 201–12.

166L During flower or fruit shortages, many vertebrates increase the importance of other food groups in their diets or resort to inferior foods, like bark or leaves: Mark Leighton and Dee Robbins Leighton, "Vertebrate Responses to Fruiting Seasonality within a Bornean Rain Forest," in *Tropical Rain Forest: Ecology and Management*, edited by Stephen L. Sutton, Timothy C. Whitmore, and A. C. Chadwick (Oxford: Blackwell Scientific Publications, 1983), 181–96.

166L Continual flowering by many understory plants has been shown by Linda E. Newstrom, Gordon W. Frankie, and Herbert G. Baker, "Survey of Long-Term Flowering Patterns in Lowland Tropical Rain Forest Trees at La Selva, Costa Rica," in *L'Arbre, Biologie et Développement*, edited by C. Edelin (Montpellier: Naturalia Monspeliensia, 1991), 345–66.

166R John Terborgh, "Keystone Plant Resources in the Tropical Forest" in *Conservation Biology: The Science of Scarcity and Diversity*, edited by Michael E. Soulé (Sunderland, Mass.: Sinauer Associates, 1986), 330–44; Annie Gautier-Hion and Georges Michaloud, "Are Figs Always Keystone Resources for Tropical Frugivous Vertebrates? A Test in Gabon," *Ecology* 70 (1989), 1826–33. (The answer to the question is no, at least not in Africa.)

CHAPTER 10: *A Science Nears Maturity*

172L Use of the canopy sled to sample insects has been described by Margaret D. Lowman, Mark W. Moffett, and H. Bruce Rinker, "Insect Sampling in Forest Canopies: A New Method," *Selbyana*, in press.

173L The canopy raft expedition to French Guiana is described by Francis Hallé, "A Raft Atop the Rain Forest," *National Geographic* 178 (October 1990), 129–38.

174R Francis Hallé discusses his recent views about trees in "Le Bois Constituant un Tronc Peut-Il être de Nature Racinaire? Une Hypothèse," in *L'Arbre, Biologie et Développement*, edited by C. Edelin (Montpellier: Naturalia Monspeliensia, 1991), 97–112.

174R Fusing of individuals in fig trees is described by James D. Thomson, Edward Alan Herre, James L. Hamrick, and J. L. Stone, "Genetic Mosaics in Strangler Fig Trees: Implications for Tropical Conservation," *Science* 254 (1991), 1214–16. Other aspects of the uniqueness of figs are discussed by Daniel H. Janzen, "How to Be a Fig," *Annual Review of Ecology and Systematics* 10 (1979), 13–51.

176L Bradley C. Bennett, "Uses of Epiphytes, Lianas, and Parasites by the Shuar People of Amazonian Ecuador," *Selbyana* 13 (1992), 99–104; and "Plants and People of the Amazonian Rainforests," *BioScience* 42 (September 1992), 599–607.

176L John P. Dumbacher, Bruce M. Bechler, Thomas F. Spande, H. Martin Garaffo, and John W. Daly, "Homobatrachotoxin in the Genus *Pitohui*: Chemical Defense in Birds?" *Science* 258 (1992), 799–801.

ACKNOWLEDGMENTS

Quote: John Steinbeck, *The Log from the* Sea of Cortez (New York: Viking Press, 1951).

INDEX

Illustrations are indicated by italicized page numbers

adaptations for climbing: in animals, 124–126; in plants, *84–85,* 86–88
African tropics, 9, 19, 59, 103, 105, 113, 120, 124, 129, 131, 139, 150, 161. *See also* Cameroon; Congo basin; Uganda
air plants, *70,* 71, 72. *See also* epiphytes
algae, 69, 73, 124
Amazon basin, 17, 24, 30, 35, 38, *49,* 51, 58, 80, 101, 107, 159–160, 176
American tropics, 18–19, 41, 71, 81, 94, 109, 110, 115, 131, 139, 144, 147, 150, 161, 166. *See also* Amazon basin; Andes; *individual country names*
Anacardium trees, *42,* 118, 119
Andes, 73, 81, 90, *116,* 118–120
animal-plant interactions, 113, 136. *See also* ant plants; fruit dispersal; herbivory; mites; pollination; seed predators
Anolis lizards, 122, 126
ant gardens, 99, 113, 114, 161
anthuriums, 77
ant plants, *44,* 99, 112–113, 114
ants, *15, 99,* 110–115, *115,* 118–119, *145,* 161
Appanah, Simmathiri, 136–139, 142–143
aquatic life in canopy, *74,* 109
architecture. *See* crowns; stratification
Aristolochia vines, *98*
army ants, 111
Ashton, Peter, 136, 138, 150
Asian tropics, 16, 19, 27–31, 35, 41–43, 62, 81, 131, 132, 139, 143, 144, 145, 146, 147–150, 154, 156, 160, 161, 164, 166. *See also* Borneo; *individual country names*
Australia, 30, 31, 52, 81, 105, 126, 131, 175
Azteca ants, 112, 113, 114

balsa trees, 44, *144*
bamboos, 48, 86
Barro Colorado Island, 12, 37, 41, *42,* 84, 86, 164, 166
Basset, Yves, 109
Bates, Henry Walter, 35, 58
Bates, Marston, 23
bats, 131, 132, 147, 168; as frugivores, 164, 168; as pollinators, 142, 144, 145, 168. *See also* flying foxes
Beaman, John, 28, 29
bears, spectacled, *116,* 117, 120, 125
Beebe, William, 25, 101
bees, 21, 110, 139, 142–147, 166
beetles, 24, *24,* 102–103, 105, 107, 108, 124
Belt, Thomas, 109, 122
Benzing, David, 73, 80
Berner, Pierre, *46,* 47–50, 54–55
biodiversity, 25, 101–102, 104–106, 115, *119,* 150, 154, 156, 157
biogeography, 35, *45*
Biological Dynamics of Forest Fragments project, 133
birds, *117,* 118, 148, 150, *153*; as frugivores, 162, 163; as pollinators, 144–145, 161; flight, 131–132, 147. *See also* hummingbirds
Block, Carl, 28
booms, canopy, 136–137, *137*
Borneo, *2,* 16, *21, 26,* 27–29, *27,* 30–34, *30,* 35, 72, *75,* 81, 83–84, 104, 122, 130, *152, 153,* 154, *160,* 161–163, *162,* 166–168, *169*
botflies, 106
brachiators, 121, 125
Braulio Carrillo National Park, 66
Brazil, 23, *32–33,* 51, 132–*133*
Brockelman, Warren, 121
bromeliads, *65,* 66, *70,* 71, 73, 74, *74,* 76, *76,* 77, 80, 81, 109, *116,* 117, 125, 142, *142. See also* tanks
Brown, William L., Jr., 112
Browne, Frank G., 148
bullet ants, 114
bumblebees, 146, 147
butterflies, 98, 104, 144, 147, *149,* 158

Cameroon, *76,* 118, *170,* 171–175, *173*
camouflage, 107, *107*
CAM plants, 77
Camponotus ants, 99, 113
Cannonball tree, *166*
canopy, definition of, 24; biology, as profession, 22–25
canopy structure. *See* crowns; crown shyness; gaps; stratification
carnivorous plants, 109
carpenter bees, 144
Carr, Archie, 131–132
cauliflory, 164, *165, 166*
Cecropia, 32–33, 44, 94, 113
chameleons, 126, 128, 131
Chapman, Colin, 121
Chapman, Frank, 51
Chapman, Lauren, 121
Charles-Dominique, Pierre, 118
chemical signals, 114, 124
chimpanzees, 121, 125, *129*
Chinnappa, K. M., 129
Chloropid flies, *111*
Clark, Ken, 71, *71,* 72, 77
claws, climbing with, 125
Cleyet-Marrel, Dany, 174
climax trees, 31, *32,* 41, 44, 52, 105, 133, 160, 182n41R
climbing, by animals, 62, 107, 113–114, 117–122, 124–126, *125,* 128–129, *129,* 131, 133, 159, 160, 164
climbing, by humans, 22–24; bare handed, *30, 92;* cherry pickers, 24; ladders, *136,* 146; *pecohna,* 132–133, *133;* rope climbing, 21–24, 47, 49, 50; spiked shoes, 23. *See also* booms; cranes; rafts; sleds; towers; walkways
climbing plants (vines and lianas), 19, 43, 62, 66, *82,* 83, 84, *84, 85,* 86–89, *87,* 90, 93, 94, 98, 147, 163
cloud forests, 19–20, *20, 29, 40,* 48, *64,* 73, 81. *See also* La Planada Reserve; Monteverde Cloud Forest Reserve
Clusia, 91, 93, *93,* 97
colobus monkeys, 121, 124
Colombia, *67, 74, 76, 79,* 80, 89, *89,* 90, *90, 91,* 93, *93,* 112, *115, 116,* 117, *118, 135*
coloration: in animals, *107, 119,* 123; in plants, 53, 73
communication, 114, 122–124
competition, 36, 38, 43, 44, 54, 69, 78, 94, 112, 113, 126, 132, 145–146, 147, 174
Congo basin, 31–32, 34, 35, 89, 104, 157
conservation, 9, 25, 77, 176, 181
coqui frogs, 130, 131
Corcovado National Park, *82*
Corner, Edred, 51
Costa Rica, *14,* 16–17, 21–22, *23, 24, 44, 46,* 47–50, 53–55, *60–61,* 62–63, 66–69, *69, 70, 71,* 71–73, *82, 92,* 107, *119,* 122, *123, 140–141, 142, 167*
Coussapoa hemiepiphytes, *95*
Coxson, Darwyn S., 80
crabs, 109
cranes, canopy, 53, 54, 81, 175
Croat, Tom, 159
crowns, tree: architecture of, *5,* 29, *29, 44, 47,* 51, 54–55, 58, 80, 89, 105, 129, 160, 174; shape of, 19, *29,* 30, 58, 59
crown shyness, 19, 35–36, 55, *56–57,* 59, 62, 88
cuipo trees, 156
cuscuses, 126, 131

Daceton armigerum, 111, 112, 114
dangers, of climbing, 17, 21–22, 62–63, 120, 122, *127*
Davidar, Priya, 161
Davidson, Diane, 113
dehydration. *See* water stress
Dial, Roman, 24, 122
dipterocarps, 30–31, *34,* 35, 38, 131, *138,* 150; mass flowering, 139, 145, 147, 149–150; nectaries, 114; pollination, 136, 138–139, 142, 145, 146; seed dispersal, 38, 147–150. See also *Shorea*
dirigible, hot-air, *170,* 171, 172, 173, 174
diseases, of plants, 36–38, 105, 149, 160, 182n36R
disturbance and diversity, 43–44, *119,* 133
diversity. *See* biodiversity
Dobzhansky, Theodosius, 51–52
Dodson, Calaway H., 73
Doliocarpus lianas, 87
Draco lizards, 131
dragonflies, 109
durians, 162–163, 164

Ebersolt, Gilles, 172, 174
Ecuador, *18–19,* 31, 34, *48,* 73, *74, 87, 95*
Elaeocarpus amoenus tree, *148–149*
elfin forests, 20, *21,* 73
emergent trees, 41, 53, 59, 147, 150, 160, 163
Endler, John, 123
epiphylls, 77
epiphytes, 66, 67, 68–69, 70, 71, 72–73, 74–75, 76–77, 80–81; communities, 77–78, 80; seed dispersal, 161, 163. *See also* orchids
Erwin, Terry, 100, 101, 102–103, 104, 105, 111, *180*
euphonias, 161
evolution, 35, 97, 181

famine and food reliability, 139, 142, 145, 149–150, 164, 166
Farrell, Brian, 105
ferns, *47,* 48, *74,* 77, *81,* 86
figs, *60–61,* 63, *70, 92,* 145–146, *145,* 152, *153,* 164, *165,* 166, 167, 174; hemiepiphytes, *92,* 93–94, 166, 167, 174
fig wasps, 145, *145,* 146, 157, 166
fireflies, 110
Fischer, Keyt, 126, *127*
fish, 159–160
flies, 106, *111,* 142
flocking, in birds, 118
flooded forests, 35, 101, 107, 159–160
flowering syndromes, 144, *148–149,* 157–158
flowerpeckers, 161
flying animals, 108, 122, 132, 138–139, 142, 144–145, 147, 159, 163, 164
flying dragons, 131
flying foxes, 147, 150, 160, *168*

flying snakes, 130, 131
flying squirrels, 130, 131
flyways, 62, 132, 163, 164
fogging, arthropod, 101, 103, 105, 110, 112
Foster, Robin, *11*, 37, 41, 43, 164
French Guiana, 173
frogs, *17*, 74, 109, 126, 130, 131, 176
frugivores, 105, 154, 160–164, 168, 188n154R
fruit, *93*, 154, 159, 188n154R
fruit dispersal, 113, 154–168. *See also* seed dispersal
fruiting syndromes, 159, 161, *162*, 163–164
fungi, 37, 38, 73, *74*, 75, 77, 78, 113, 115; plant roots, 38, 43, 44, 70, 77, 159

gaps, canopy: bridged by climbing animals, *17*, 62, 114, 118–122, 125, *125*, 128–129; bridged by gliding animals, 130–131; bridged by plants, 88, 89, 96; treefalls, 38, *39*, 40–41, 43, 55, 62, 80, *85*, 89, 97, 133, 144, 147. *See also* crown shyness
gene dispersal, 155–157, *168*
Gentry, Alwyn H., 73
giant flying squirrel, 130
giant squirrels, 128, *151*
gibbons, 19, 121, 122, 128
Gilbertiodendron trees, 31, 34, 35, 38, 89
gliding animals, 130–131
Goulding, Michael, 107
gravity, issue in canopy biology, 104, 107, 120–122, 125
Guam, *43*
Gunatilleke, Nimal, 146
Gunatilleke, Savi, 136, 146
Gunung Palung National Park, 154, 161–164, 166–169
gympie trees, 52, 105

Hallé, Francis, 174
Hamrick, Jim, 156
Hansen, Eric, 72
Hawaii, *45*
hawkmoths, 158
height, human response to, 62–63
Heliconius butterflies, 104, 147
helicopter damselflies, 108, 109
hemiepiphytes: figs, *92*, 93–94, 166, 167, 174; primary, 66, *82*, 89, 90, *90*, 93–94; secondary, 94, 96–97
herbivory, 36–38, 104–106, 114, 115, 160, 175, 182n36R, 185n105L
Hercules beetles, 103
Herrera, José, 53
Herwitz, Stanley, 81
Hogue, Charles, 53
Honduras, 131–132
honeybees, 146, 147
hornbills, 132, 150, 162, 163
Hubbell, Stephen, 37
Hudson, W. H., 5
Humboldt, Alexander von, 25
humidity, 35, 68, 77
hummingbirds, 132, 136, 142, 144–145, *155*, 163
humus. *See* soil, canopy

iguanas, *120*
India, 20, 31, 112, *125*, 129–130, *151*, *157*, 161
Indonesia, *2*, *30*, 100, 112, 122, 131, *152*, 153–154, *153*, *160*, 161–163, *162*, 166–168, *169*
insects, 36–37, 99–115, 122, *124*, 135, *148–149*, 158, 159, 172. *See also specific groups*

jumping. *See* leaping animals
jumping spiders, *6–7*, *15*, 16, *102–103*, *107*
jungles, 43–44, 88

kapok trees, 41, 59, 118, 119, 120
kapur trees, *56–57*, 62
Kazda, Marian, 172
keystone species, 164, 166–168
Khao Yai National Park, 121
Kibale Forest, 121, 129
koa forests, *45*

Laman, Timothy, *2*, 166–*169*
Langtimm, Cathy, 122, *123*
langurs, 157, *157*
La Planada Reserve, 89, 90, 93
leafcutting ants, 115, *115*, 118–119
leafhoppers, 142, 143
leaping animals, 121–122, 125, 128–129
leaves: and light, 54; and wind, 63; drip tips, 52, 77; flushes, 54, 105, 115, 185n105L; shape, 52–53, 73, 96, *96*; turnover, 53–54, 76, 105, 147, 160
Leighton, Mark, 154, 162, 163
lemurs, 131, 159
lianas, definition, 86. *See also* climbing plants
lichens, 73, *75*, 77, 78
light: and stratification, *39*, 54–55, 58–59, 60, 68, 72, 77, 78, 97, 112, 133; and seedling survival, 36, 40–41, *42*, 44, 93, 94, *96*, 167
liverworts, 73, 77
lizards, *120*, 122, 126, 128, 131
Longino, Jack, 17, 21, 22, *23*, *60*, 63, 78, 107
lowland rainforests, *18–19*, 19, 21, 30, 112
Lowman, Margaret, *173*, 175
Lugo, Ariel, 80
Luquillo Mountains, 122, 130
Luvall, Jeff, 66, 67, 68, 73
Lycopodium epiphytes, *75*

macaques, 124, *125*, 128, *152*
Machado, José Luis, *42*
Madagascar, 9, 112, 158, 159
Malaysia, *21*, *26*, 27, *27*, 28, 29, 30, 35, *39*, *56–57*, 62, *75*, 111, 122, 130, 136–139, *137*, *138*, 142–143, 146, 149, 162
Malcolm, Jay, 132–133, *133*
mammals: and canopy conditions, 128; size in trees, 128–129. *See also* vertebrates
mangrove trees, 35, 62
Mares, Michael, 101
masting, 38, 148–149, 154, 164
Matelson, Teri, 69, *70*, 72, 78
Matthiessen, Peter, 88
May, Robert, 100
McClure, Elliot, 111, 122
Mena, Elieser, 50
mice, 119, 122, 123, 125
Microcos trees, 163
mimicry, *15*, *107*, 143
mist, 68, 71–72, 73, 77
mistletoes, 66, 73, 89, *89*, 90, *90*, 93, 161
Mitani, John, 122
mites, 108, 124; plant interactions, *106*, 113
monkey-ladders, 86
monkeys, 118, 119, 121, 124, *125*, 128, 131, 133, *152*, *157*, 162, 163, 164, *167*
monodominant (low-diversity) forests, 31, 34, *34*, 35, 38, 43, 44, *45*, 48, *56–57*, 59, 89, 104, 139, 182n43R
monsteras, 94, 96, *96*
Monteverde Cloud Forest Reserve, 17, 21–22, *24*, 25, *60–61*, 63, 68–78, *70*, 71–73, *71*, 76–78, *92*, 107, 122
mosquitoes, 65, 106
mosses, 73, 77, 78
moss forests, 22, *22*, *69*, *75*
moths, 105, *107*, *124*, 144, 158
mouse opossums, 119
Murawski, Darlyne, 84, 86, 87, 88, 89, 155, 156, 157, *180*
Muul, Illar, 28
Myers, Norman, 30

Nadkarni, Nalini, 22, *23*, 68–69, *69*, 70, 71, 72, 77, 78, 80, 107, 174
Nagara Hole Reserve, 129–130
naked Indian tree, *79*, 80
nectaries, 114–115, *144*
New Guinea, *20*, *22*, *29*, 30, *40*, *47*, *69*, *74*, *75*, *81*, *98*, *99*, *106*, *107*, 126, *127*, 128, 131, *139*, *145*, *158*, *168*, 175–176, *177*
Ng, Francis, 55, 58
Nicaragua, 109
Nickerson, Max, 16
Nilgiri Hills, 112, 161
nocturnal mammals, 119, 122–123
nutrient cycling, *70*, 72, 78, 81

oaks, *46*, 48, 55, 142
oilbirds, 161
old growth forest species. *See* climax trees
Opération Canopée, 172–175
opossums, 119, *119*, 122, 128, 131, 133
orangutans, 120–121, 129, 153, 154, 161–162
orchid bees, 143–144, 146
orchids, 67, *67*, *75*, 76, 77, 78, 143–144, 158, 159
Oryctanthus mistletoes, 89

Pacaya-Samiria Reserve, 101
palm trees *32–33*, 80, 83, 84, 94, *95*, 131, *169*
Panama, *10–11*, *17*, 37, *37*, *38*, *42*, *52–53*, 54, 81, 84, *84*, *85*, 86–88, *96*, *107*, 109, *120*, *124*, *144*, 155, *155*, 156, 164, *171*, *180*
Pandanus trees, *127*, *169*
pangolins, 131
Paoletti, Maurizio, 178
parachuting animals, 130, 131
parasitic animals, 106
parasitic plants. *See* mistletoes
parrot snakes, 16, *17*
pathogens. *See* diseases
Paso forest, 137, 139
Passiflora vines, 163
pecohna, 132–133, *133*
peepul tree, 94
Perry, Donald, 23–24, 40
Peru, *49*, 99, 101, *180*
Philippine mahogany trees, 136, 139, 142, 143, 145, 148
Philippines, 30
philodendrons, 85, 94
photosynthesis, 59, 77, 90
physical environment, and trees, 35–36. *See also* temperature stress; water stress
pigeons, green, 148, *153*
pioneer trees, 31, *32–33*, 41, 43, 44, 48, 53, 55, 133, 160, 174, 182n41R
pitahoui birds, 176
plankton, aerial, 25, 108, 122, 142, 159
Platypodium elegans, 37, *37*, 41
poison dart frogs, 109, 176
pollination, 136, 138–139; and stratification, 146–147, 163–164; mass flowering, 139, 142; payoffs for pollinators, 143–145, 161; pollen dispersal, 138–139, 144–146, 154–159, 164; sequential flowering, 139, 142–143. *See also* flowering syndromes
population genetics, 155–157
Poring Hot Springs, *26*, *27*, 28
posture, in climbing animals, 125
pottos, 118, 128
praying mantis, *171*
predators, 106, *107*, 187nn128–129. *See also* seed predators
prehensile tails, 124, 128, 131
prescription drugs, 104, 176
primates, 19, 118, 120, 121, 122, 128, 131, 159, 162, 163. *See also* monkeys; orangutans
profile diagrams, 175
Pseudobombax septenatum trees, *144*

INDEX

Psiguria lianas, 147
Puerto Rico, 24, 122, 130
Putz, Jack, 86, 88, 94

Raciborski, M., 84
rafts, canopy, *172–173*, 173–175
rainfall: reproduction cue, 164; effects on canopy, 35, 67, 68, 73, 80–81. *See also* mist
rainforests, tropical: definition of, 20–21; by elevation, 19–20; regional differences, 18–21, 80, 81, 131, 139, 150
rarity: of insects, 100, 101; of trees, 31–34
rattans, 83–84
Ray, Tom, 94
red-eyed tree frogs, *17*
red jungle fowl, 160
refugia, 35, 182n35L
reproduction, plant, 154–155. *See also* pollination; seed dispersal
Richards, Paul W., 9
Ridley, Henry N., 148
Rinker, Bruce, *173*, 175
roots: of climbing plants, *85*, 87; of epiphytes, 67, *69*, 70–71, 72, 77, 78; of mistletoes, 90; of primary hemiepiphytes, 93–94; of secondary hemiepiphytes, 96–97, *96*; of trees in crown, *69*, 70–72, 80, 174
rope climbing, 21–24, 47, 49, 50
roughness, canopy, 20, 63, 81
routes, through canopy, 114, 118–120, 131. *See also* flyways
rubber trees, 38, 105, 159–160

salamanders, 125–126
Samper, Cristián, 89, 90, *91*, 93
sandbox tree, 159
Sanderson, Ivan, 121
Savant, Grace, 178
scale insects, 115
Scatina, Frederick N., 80
Schmidt, Karl, 131
Schweinfurth, Georg, 58
scrambling plants, *85*, 87, 88
screw palm, *169*
secondary dispersal, 160–161
seed dispersal, 147–150, 154–156. *See also* fruit dispersal
seed predators, 36–38, 148–150, *153*, 154, 157, 160
seed shadows, 36, 37, 160
Shelford, Robert, 130
Shorea, 30, *34*, 59, *134*, *137*, *150*. *See also* Philippine mahogany trees
Shoumatoff, Alex, 157
siamang, 122
Singapore, 162, *165*
Sinharaja Biosphere Reserve, 146
Skutch, Alexander S., 5
sleds, canopy, *170*, 171–174
sloths, *124*, 125, 126
Smith, Alan P., 54, 81
Smithsonian Tropical Research Institute, 37, 53, 54, 81
snakes, 16, *17*, *118*, 119, 128, 130, 131
Soberania National Park, 155
soil, canopy: and plants, 68–72; and animals, 106–108
Southwood, Sir Richard, 104
Spanish moss, *65*, 66, 73
species: new to science, 100–101, 107, 176; number of, 100–102
spider monkeys, 118, 119, 121, 124, 164, *167*
spiders, *6–7*, *15*, 16, 102, *102–103*, 107, *107*
Spruce, Richard, 17
squirrels, 128, 129, 130, 131, *151*, 160
Sri Lanka, *6–7*, *15*, 17, 31, *34*, *75*, *102*, *107*, *134*, 146, *148–149*, *150*
Stanley, Henry Morton, 105
Star Mountains, 175
Stelis orchid, *67*
Sterculia fruits, *158*
Stewart, Margaret, 130
Stork, Nigel, 109, 112
storms, 40, 41, 66
stranglers, *92*, 93–94, *95*, 97, 174, 184n94L
stratification, forest, 9, 20, *49*, 51, 58–59, 105, 112, 163–164, 166, 183nn59L,59R; and body size of mammals, 128–129; and plant reproduction, 146–147
stress. *See* temperature stress; water stress
succession, 41, 55. *See also* climax trees; pioneer trees
suction, climbing with, 125–126
Swain, Roger, 23

Tabebuia rosea trees, *167*
Talamanca Range, *46*, 48–49, 54–55
tamarins, 122
tambaqui fish, 160
tanks, bromeliad, 74, 77, 80, 109, 125
temperature: cue to reproduce, 143, 150; rainforest, 68
temperature stress: animals, 68; plants, 68, 72–73, 76, 78, 86
tendril-bearing plants, *85*, 86, 87, 88
Terminalia tree, *157*
termites, 110
Thailand, 35, 110, 121
thrips, 135, 136, 138–139, *138*, 142, 143, 145
Tobin, John, 17, 21, 66, *180*
Tomlinson, Henry, 24
toucanets, *117*, 118
toucans, 132, 150
tourism, 28
towers, climbing, 65, 66, 67
toxic animals, *107*, 176
toxic plants, 52, 104–105, 113, 115, 138, 150, *157*, 163
tree ferns, *47*, 48
tree frogs, *17*, 109, 126
tree kangaroos, 126
trees: adult mortality, 40, 41, *85*, 89; bark, 51, *79*, 80, 88; buttresses, *48*, 51; community biology, 30–44; dispersion, 31–32, 36–37, 43, 59, 105, 145, 155–157; distribution, 30–31; hollow trunks, 110, 119, 186n119R; juvenile mortality, 36–38, 105, 149, 160, 182n36R; leaf shape, 51–52, 77; leafing pattern, 53–54, 105, 147, 160; population genetics, 155–157. *See also* crowns
Trinidad, 31, 104, 106
twining plants, *84*, 87

Uganda, 31, 65–66, 67, 121, *129*

value: of biodiversity, 25, 176; of products, 77, 104, 176
Venezuela, 17, 22, *111*, 112, 161
vertebrates, body size, 128–129, 150; stratification, 128–129, 133. *See also* climbing, by animals; flying animals; gaps; gliding animals; *specific groups*
vibrations and predation, 126–127
vines, definition, 86. *See also* climbing plants
Virola trees, 55
Vochyseia ferruginea, *140–141*

walkways, *26*, *27*, 28
Wallace, Alfred R., 30, 31, 88
Wallace's flying frog, 130, 131
Wallace's Line, 30
wasps, 21, 110, 172. *See also* fig wasps
water, collected by plants, *65*, 73–77, *74*, 109, 143. *See also* mist; water stress
water stress: on animals, 67, 68, 112, 128; on plants, 67, *67*, 68, 72, 73, 76–77, 78, 81, 93, 97, 112
Waterton, Charles, 125
weaver ants, *15*, *99*, 100, 112, 113–114, *145*
Williams, John, 23
Wilson, Edward O., 12, *180*
wind: effects on canopy surface, 63; in seed dispersal, 37, *37*, 147–148, 159, 163; in tree mortality, 40, 89. *See also* air plants
Wolda, Henk, 109
Wolf, Jan, 178
woodnymphs, crowned, *155*

Yonkow, Niki, 162

Zens, Scot, 163
Zotz, Gerhard, 178, *180*

Designed by Amanda Wilson

Composed in Caslon 540, Caslon 3, and Copperplate 33BC with QuarkXpress 3.1 on a Macintosh IIsi by Barbara Sturman

Printed and bound by Grafiche Milani, Milan, Italy

Production by Hope Koturo and John F. Walsh